VINODIVERSITY: THE BOOK

New Varieties and New Wines in Australia

Darby Higgs

ABOUT THE AUTHOR

Darby Higgs is a wine writer from Melbourne, Australia. He is the founder and owner of *Vinodiversity.com* the most complete source of information about wine varieties in Australia.

This edition Published 2010
Copyright © 2010 by Darby Higgs

An earlier version of the material published in this book was previously published as *Emerging Varietal Wines of Australia*. It has been extensively expanded and updated.

All rights reserved. Other than brief extracts for the purpose of review no part of this publication may be reproduced in any form without the written consent of the copyright owner.

The author has done his best to ensure the accuracy and completeness of this guide. However, he can accept no responsibility for any loss, injury, or inconvenience sustained as a result of information or advice in this guide.

Contents

Introduction 7

How to use this book 11

Varieties 13

Regions 101

Wineries 133

References 199

Introduction

WHAT IS VINODIVERSITY?

Wine is the most diverse product that we can buy. There are literally thousands of brand names of Australian wine. These brands are owned by huge international corporations, middle sized companies down to small single families or individual makers.

The whole point of wine is that each variety, each style, each region, each vintage is different. But there is a paradox, with so much choice and diversity consumers opt to buy mostly mass marketed wines from a narrow range of varieties. Winemakers struggle to sell anything that is slightly off beat. Wines sell on fashion rather than on quality.

In Australia wine is made from about 140 different grape varieties. However few wine consumers, even those who I meet at wine shows, can name more than a couple of dozen. We have a treasure trove of different, exciting and generally good to excellent wines made from unusual, or alternative varieties that deserve at least a try.

HOW MANY WINE GRAPE VARIETIES ARE THERE?

This is a question without a definitive answer. It could be around 5,000 or it might be 10,000 depending on your definition of variety. Do you count the variety that has yet to be released from the plant breeding station? Or what about the variety that is represented by just a few vines in an obscure corner of a little known region?

CONFUSING NAMES

Many varieties have different names in different parts of the world. It is only in recent years with increasing emphasis on labels for international trade and the application of DNA sequencing that some varieties have been reliably identified.

WHERE DID THESE VARIETIES COME FROM?

Most agricultural plants and animals exist as different varieties that have arisen by the combined forces of natural selection and deliberate breeding and propagation practices of farmers and viticulturalists.

The majority of older wine grape varieties arose several centuries ago as a result of grape growers electing the best vines and propagating from them via cuttings or grafting. Gradually more scientific approaches were used, first by individuals and later by large research institutes.

A major boost to grape breeding occurred as a result of the *Phylloxera* pest which invaded many vineyards during the last half of the nineteenth century. Many varieties were bred by crossing American vines with European vines to overcome this pest and other diseases.

All grapes belong to the Vitis genus, of which there are about 60 species. European vines are members of the *Vitis vinifera* species. A few North American species such as *V labrusca* and *V rupestris* are also important because of their resistance to *Phylloxera* and to some fungal diseases.

BLENDED AND VARIETAL WINES

Wines are either varietal or blends. Strictly speaking a varietal wine is made from 100% of a single grape variety. However the labeling laws in most countries allow small amounts of another variety or varieties (up 10 or 15 %) to be added to the wine, so varietal pure wines are a little less common than you might think.

Blended wines are composed of two or more grape varieties and in New World wines these are disclosed on the label with the predominant variety named first. Thus a common South Australian blend is GSM a blend of three different red grape varieties Grenache, Shiraz and Mourvedre.

Sometimes blended wines are made by mixing varietal wines which have been made separately; sometimes a mixture of grapes is fermented together.

A special type of blending is called co-pigmentation. The best known example known is Shiraz Viognier. In this case a small amount of a white grape variety, Viognier, is blended with Shiraz at the fermentation stage. During fermentation a different biochemical pathway occurs with the natural pigments and flavours in the grape skins. The result is that the wine develops a brighter and more stable color and subtle changes to the flavour of the wine occurs.

THE WINE CENTURY CLUB

People learn about wine varieties for all sorts of reasons, to plant grapes, to make wine, or to find new and exciting wines to enjoy. One group of people learns about wine varieties so they can become members of the Wine Century Club. We are a small but growing group of people throughout the world who have tasted over a hundred varieties and are interested in pushing the boundaries and finding new varieties to try and then counting them.

Find out more at http://www.vinodiversity.com/winecentury

CORRECTIONS AND ADDITONS

Up to date information about the topics covered in this book can be found at

www.vinodiversity.com

You can use the contact page on that website to forward errors and additions to the author.

Vinodiversity.com also contains information about all of the varieties and wine regions discussed in this book as well as articles on a range of wine related topics.

How to use this book

This book is about wines made from **alternative** grape varieties. It does not seek to cover the mainstream varieties of Cabernet Sauvignon, Chardonnay, Merlot, Pinot Noir, Riesling Sauvignon blanc, Semillon or Shiraz. The coverage of Cabernet franc is limited to its use as a varietal wine. It is used more commonly in blends, but that is outside the scope of this book.

The first section - VARIETIES - lists grape varieties that are used to make wine in Australia. Most entries includes include some references from which the information is drawn, listed in code form.

> **C - Clarke, Oz & Margaret Rand** *Grapes and Vines: a Comprehensive Guide to Varieties and Flavours* (London: Websters International Publishers 2003
> **D - Delong, Steve & Deborah De Long** *Wine Grape Varietal Table: Wine and Grape Indexes* (London: De Long Wine Info 2004)
> **H - Halliday, James** *Varietal Wines (*Sydney: HarperCollins 2004)
> **K.- Kerridge, George and Allan Antcliff** *Wine Grape Varieties Revised edition (*Melbourne : CSIRO 1999)
> **O - Robinson, Jancis** *Oxford Companion to Wine, 3rd edition (*Oxford: Oxford University Press 2006)

The REGIONS section lists and briefly describes the various GI Zones Regions and Subregions in Australia. They are arranged into chapters by state. Each of the regions entries is concluded by a list of the wineries which are using alternative varieties and which can be found in the WINERIES index which concludes this book.

The WINERIES list contains all of the wineries listed in the Vinodiversity database of alternative varieties. Web addresses are included where known. Wineries are constantly being opened, taken over, closed or changing names. The list is therefore always incomplete and out of date, but it represents a fair sample of the wineries that were operating in April 2010 or in the previous year or two.

Varieties

The varieties listed below can be loosely described as 'alternative'. Some are quite common but many Australian wine consumers are unfamiliar with them.

Each entry includes a list of wineries which are using the variety.

1893

Wine type: Dry white

References: This variety is a one off so it doesn't appear in standard references. See Rimfire vineyards website at http://www.rimfire.com.au

This is really a variety without a name; in fact it is an unidentified white grape variety, a relic from an historic vineyard planted in 1893. This vine's DNA is not known anywhere else in the world. Rimfire vineyards in Queensland are thus the only known producer.

- Rimfire Vineyards (Queensland)

AGLIANICO

Maturity group: Late ripening variety, Wine type: full bodied, high acid red

References: C, D, H, O

Aglianico is a highly regarded red wine variety from the south of Italy. As more Australian winemakers become concerned about climate change they are turning to varieties that thrive in warmer conditions and this is one that has attracted attention. The wines are typically well flavoured, medium to full bodied and have a high acidity even in warmer climates.

- Brown Brothers (King Valley) Chalmers (Murray Darling) Di Lusso Estate (Mudgee) Karanto Vineyards (Langhorne Creek) Pertaringa (McLaren Vale) Rimfire Vineyards (Darling Downs) Sutton Grange Winery (Bendigo)

ALBARINO

Synonyms: Alvarinho, Albarin Blanco,

Maturity group: Late ripening variety, Wine Type: Light bodied wines with high acidity.

References: C, D, O

Early in 2009 it was determined that the vines planted in Australia which were believed to be Albarino are in fact Savagnin. Wines produced from those vines are now correctly called Savagnin and are dealt with in the entry under that name in this chapter. Most of those who thought they had Albarino are happy with the wine made from Savagnin.

There are now no wineries using of Albarino in Australia, but that could change. Given the growing popularity of all things Spanish I'm sure someone will take on the task of introducing true Albarino.

ALEATICO

Synonyms: Agliano, Leantico, Moscatello, Livatica

Wine Type: Very full bodied wines with low acidity.

References: D, H, K, O

Aleatico is a highly perfumed red grape variety with some similarity to Muscat. It is used in Italy and on the Mediterranean islands of Elba and Corsica to produce sweet wines, some are fortified. It is also grown in Central Asia and Chile. A handful of Australian wineries, mainly in the warmer regions are using Aleatico.

- Di Lusso Estate (Mudgee) Freeman Vineyards (Hilltops) Hollyclare (Hunter) Rimfire Vineyards (Darling Downs) Riversands Winery (Queensland Zone) Tizzana Winery (South Coast Zone)

ALICANTE BOUSCHET

Synonyms: Alicante Henri Bouschet, Garnacha tintorera

Wine Type: Very full bodied wines with low acidity.

References: C, D, H, O

This red wine variety is falling from favour in most areas of the world where it is grown. Curiously, unlike most other varieties the juice as well as the skin is pigmented. Varieties with coloured juice are called 'teinturier' and they were used to add colour to otherwise pale red wine - a useful factor in the days of mass produced vin ordinaire. Alicante Bouschet is the earliest ripening variety in Southern France. It is designated by Jancis Robinson as a "workhorse variety" but these days variety is unpopular because the wines made from it tend to lack structure.

The variety was produced by Henri Bouschet in the mid 19th Century as a crossing with Grenache. In fact 'Alicante' is one of the many synonyms for Grenache, but Alicante Bouschet is a distinct variety.

The variety is used by just a few wineries in Australia, mostly as a component of red blends. Rockford in the Barossa Valley makes a well regarded rose wine from Alicante Bouschet.

- Diggers Bluff (Barossa) Forester Estate (Margaret River) Loan Wines (Barossa) Rockford (Barossa) Taminick Cellars (Glenrowan) Vinden Estate (Hunter) Virgara Wines (Adelaide Plains)

ALIGOTE

Synonyms: Chaudenet Gras, Blanc de Troyes, Plant Gris

Maturity group: Early maturing, Wine Type: Light bodied wines with high acidity.

References: D, H, O

Aligote is the second string white grape variety (after Chardonnay) in Burgundy. Here and in nearby Chablis it is used for basic wines intended to be drunk young or mixed with cassis (blackcurrant liqueur) to make the refreshing aperitif called kir. The only Australian user for this variety that I know about is Hickinbotham winery on the Mornington Peninsula, who blend some Aligote into some of their Chardonnay.

- Hickinbotham (Mornington Peninsula)

ARANEL

Wine type: White wine.

This variety is very obscure and none of the standard texts seem to recognise it. It was apparently bred in France and the only Australian producer seems to be Beelgara Estate.

- Beelgara Estate (Riverina)

ARNEIS

Synonyms: Barolo Bianco, Bianchetta di Alba, Bianchetto Albese

Wine Type: Light bodied wines with low acidity.

References: C, D, H, O

This elegant Italian white wine variety is creating a buzz in Australia. In its native Piedmont Arneis produces elegant white wines with powerful aromas of almonds and peaches. Arneis was much less popular in the 1970s, so much so it was near extinction, but there has been a marked increase in plantings and interest. Interestingly, Arneis has been used traditionally in Piedmont as co-pigmentation material for the red variety Nebbiolo. In this regard its history parallels that of Viognier.

Arneis presents problems in the vineyard, but better clones and handling have seen Arneis become more popular.

James Halliday credits Colin and Rosa Mitchell of Yandoit Hill Vineyard (near Castlemaine in Central Victoria) with introducing the variety into Australia.

Arneis is steadily gaining popularity in Australia. It makes pleasant light bodied white wines. In the better examples there is a pleasant aroma of herbs, honey and soft fruits.

- Bawley Vale Estate (Shoalhaven Coast) Beechtree Wines (McLaren Vale) Berrima Estate (Southern Highlands) Box Stallion (Mornington Peninsula) Brown Brothers (King Valley) Catherine vale (Hunter) Chrismont (King Valley) Coombe Farm Vineyard (Yarra Valley) Crittenden at Dromana (Mornington Peninsula) Crooked River Wines (Shoalhaven Coast) Dal Zotto Estate (King Valley) De Bortoli (Riverina) Dromana Estate (Mornington Peninsula) Dunn's Creek Winery (Mornington Peninsula) First Drop (Barossa) Geoff Hardy (McLaren Vale) Helen's Hill Estate (Yarra Valley) Kingston Estate (Riverland) Lazzar Wines (Mornington Peninsula) Mac Forbes Wines (Yarra Valley) Parish Hill Wines (Adelaide Hills) Pizzini Wines (King Valley) Port Phillip Estate (Mornington Peninsula) Rochford Wines (Yarra Valley) Rutherglen Estates (Rutherglen) Settlement Wine Company (McLaren Vale) Symphonia (King Valley) Tempus Two (Hunter) Ten Miles East (Adelaide Hills) Tertini Wines (Southern Highlands) Vale Creek Wines (Central Ranges Zone) Vale Vineyard (Mornington Peninsula) Wirra Wirra (McLaren Vale) Yacca Paddock Vineyards (Adelaide Hills) Yandoit Hill Winery (Bendigo) Yarraloch (Yarra Valley) Zonte's Footstep (Langhorne Creek)

AUCEROT

Wine type: Can be used for white wines or 'white port' style.

References: H

This wine is referred to a "mystery wine". The Ciavarella vineyard was planted from cuttings from the Bailey's vineyard in Glenrowan, which was planted from cuttings obtained from Europe in the early 1900s. The Aucerot vines in the Glenrowan vineyard were removed in the 1980s. I remember buying a couple of bottles of "Auslese Aucerot" about 1980. It was a desert style wine that aged very well.

The variety is a mystery because no one seems to know its true identity. Aucerot is not the same as the French variety Auxerrios, despite the similar name. It is quite possible that the variety no longer exists in Europe.

Currently Ciavarella uses Aucerot as a blending partner with Verdelho to make a "white port" style as well as a late harvest dessert style with Semillon.

- Ciavarella (King Valley)

BACO NOIR

References: C, D, O

This variety is a French Hybrid. It was once widely planted in France, but its main habitat now is the eastern United States and Canada. It is known for producing soft fruity wines and 'nouveau' styles in cool climates. In Australia Three Willows Vineyard in Tasmania seem to be the only producers using this variety.

- McWilliams (Riverina) Three Willows Vineyard (N Tasmania)

BARBERA

Synonyms: Barbera d'Asti, Barbera del Monferrato

Maturity group: This variety ripens in midseason. Wine Type: Full bodied wines with high acidity.

References: C, D, H, K, O

Barbera is an Italian red wine variety best known as the second most important Piedmontese variety after Nebbiolo. It is in fact the most widely planted variety in Piedmont, and is popular throughout Italy. Its role has been mainly to produce the everyday drinking wines of the region. After years of being in the shadow of Nebbiolo the variety is getting more attention from growers and winemakers and the resulting wines are much better.

It is also grown in Argentina and California, as well as Australia.

The naturally high acid levels of the grape are a beneficial characteristic of the variety, especially in warmer climates.

Barbera ripens at about the same time as Shiraz and Merlot, and a little earlier than Cabernet Sauvignon and Nebbiolo. On this basis it would seem that there are a large number of potential vineyard sites for Barbera in Australia.

VARIETIES • Barbera

The ideal terroir for the variety has not been agreed upon even its native Piedmont, so that there may be many years before we see the best wines produced in Australia.

Barbera is a variety that flies under the radar. While there is some interest in the variety, and about 90 wineries using it in Australia there doesn't seem to be much written about it.

- Aldinga Bay (McLaren Vale) Amulet Vineyard (Beechworth) Angas Vineyards (Langhorne Creek) Angullong Wines (Orange) Barrecas (Geographe) Boggy Creek Vineyards (King Valley) Boireann (Granite Belt) Bottin Wines (McLaren Vale) Broke Estate (Hunter) Broke's Promise (Hunter) Brown Brothers (King Valley) Carlei Estate (Yarra Valley) Catherine Vale Vineyard (Hunter) Catspaw Farm (Granite Belt) Centennial Vineyards (Southern Highlands) Chain of Ponds (Adelaide Hills) Chalk Hill Winery (McLaren Vale) Chrismont (King Valley) Clearview Estate Mudgee (Mudgee) Clovely Estate (South Burnett) Cobbitty Wines (South Coast Zone) Connor Park (Bendigo) Coriole (McLaren Vale) Crittenden at Dromana (Mornington Peninsula) Dal Zotto Estate (King Valley) David Hook Wines (Hunter) Di Lusso Estate (Mudgee) Donnybrook Estate (Geographe) Dromana Estate (Mornington Peninsula) Dumaresq Valley Vineyard (New England) Dunn's Creek Winery (Mornington Peninsula) Fairview Wines (Hunter) First Drop (Barossa) Gapsted (Alpine Valleys) Glenwillow Vineyard (Bendigo) Golden Grove Estate (Granite Belt) Grove Estate Wines (Hilltops) Heartland Vineyard (Hunter) Hollick Wines (Coonawarra) House of Certain Views (Hunter) Idlewild (Hunter) Jamieson Estate (Mudgee) King River Estate (King Valley) Kingston Estate (Riverland) Kyotmunga Estate (Perth Hills) La Cantina King Valley (King Valley) Lowe Family Wines (Mudgee) Mac Forbes Wines (Yarra Valley) Macquarie Grove Vineyards (Western Plains) Maglieri (McLaren Vale) Margan Family (Hunter) Massena Wines (Barossa) Massoni (Pyrenees) Michael Unwin Wines (Grampians) Michelini (Alpine Valleys) Monichino Wines (Goulburn Valley) Monument Vineyard (Central Ranges Zone) Morning Sun Vineyard (Mornington Peninsula) Mount Broke Wines (Hunter) Mount Langi Ghiran Vineyards (Grampians) Oatley Wines (Mudgee) Orchard Road (Orange) Pasut Family Wines (Murray Darling) Paul Bettio (King Valley) Pegeric (Macedon Ranges) Piggs Peake Winery (Hunter) Poet's Corner (Mudgee) Port Phillip Estate (Mornington Peninsula) Primo Estate (Adelaide Plains) Prince Hill Wines (Mudgee) Red Earth Estate (Western Plains) Redbox Perricoota (Perricoota) Riverina Estate Wines (Riverina) Rossiters (Murray Darling) Sevenhill Wines (Clare Valley) Skimstone (Mudgee) Thomson Brook Wines (Geographe) Tombstone Estate (Western Plains) Toppers Mountain (New England) Tower Estate (Hunter) Vale Creek Wines (Central Ranges Zone) Vico (Riverina) Vinea Marson (Heathcote) Warrenmang Vineyard (Pyrenees) Wild Broke Wines (Hunter) Witchmount Estate (Sunbury) Woodstock (McLaren Vale) Yandoit Hill Winery (Bendigo) Yass Valley Wines (Canberra) Zonte's Footstep (Langhorne Creek)

The codes for the References (C, D, H, K, O) are listed on page 9 of this book

BASTARDO

Synonyms: Cabernet Gros, Trousseau, Triffault, Maria Ardonna, Merenzao

Maturity group: Early maturing Wine Type: Very full bodied wines with high acidity.

References: C, D, K, O

Bastardo is a Portuguese variety that is used to make port, although even for that purpose it is not highly regarded. It is believed that some plantings of 'Touriga' in Australia are in fact Bastardo. Curiously this variety, under the name Trousseau, is used successfully in the cooler Jura region of North East France.

Although this variety is primarily a port variety it is also used as a component of red wine blends, and as varietal to make light red styles and rose wines. Bastardo grapes are aromatic and low in tannins; hence they are ideal for making softer styles of wine.

- Happs (Margaret River) Harris Organic Wines (Swan Valley) Kies Family (Barossa) Marybrook Vineyards & Winery (Margaret River) Mazza (Geographe)

BIANCONE

Synonyms: Grenache Blanc Productif

Maturity group: Among the latest ripening varieties, Wine Type: Dry or off dry white wines

References: K, H, O

This is a high yielding white variety which is used primarily for the production of wine for distillation, but under some circumstances it can make distinctive dry white. Its French synonym is quite revealing - Grenache Blanc Productif.

Wineries using this variety:

- Angoves Winery (Riverland) Lake Moodemere (Rutherglen) Mount Anakie (Geelong)

BRACHETTO

Synonyms: Braquet

Wine Type: Light Bodied, Moderate Acidity

References: C, D, O

Brachetto is found in the Piedmont area of North East Italy, and as Braquet in the nearby area around Nice in France. It is used to make semi-sweet sparkling wine and also a *passito* style red. Passito is the Italian term for a wines made with semi dried grapes.

The variety is currently being used by Pizzini in the King Valley to make a spritzy, low alcohol wine.

- Pizzini (King Valley)

CABERNET FRANC

Synonyms: Bouchet, Breton bordo

References: C, D, H, K, O

Cabernet franc is largely overshadowed by its better known cousin Cabernet sauvignon. However it is widespread in France, the cooler regions of Europe and most wine producing countries.

The most common use for Cabernet franc is as a blending partner, often minor, in so called Bordeaux blends. These wines are made from blends of Cabernet sauvignon, Merlot, Cabernet franc varieties, often with one or more of the less common Bordeaux varieties Malbec, Petit Verdot and Carmenere.

The second use for the variety is as a varietal wine, and it is for this reason Cabernet franc is included in this book as an alternative variety. Red wines from the Loire Valley, especially the Chinon Appellation are good French examples of Cabernet franc. It is also the variety responsible for the rose style wine Cabernet d'Anjou.

Varietal Cabernet francs tend to be light to medium bodied with a pronounced fruity flavour.

The list below is of wineries that make or have recently made varietal Cabernet franc wine. Many other Australian wineries use Cabernet Franc in blends.

- Bests (Grampians) Black Swan Winery (Swan Valley) Bloodwood (Orange) Brairose Estate (Margaret River) Briarose Estate (Margaret River) Bulong Estate (Yarra Valley) Byrne and Smith (McLaren Vale) Chalk Hill Winery (McLaren Vale) Chatsfield (Mount Barker) Cofield Wines (Rutherglen) Cruickshank Callatoota Wines (Hunter) Gapsted (Alpine Valleys) Goona Warra Vineyard (Sunbury) Grassy Point Coatsworth Wines (Geelong) Happs (Margaret River) Harcourt Valley (Bendigo) Hastwell and Lightfoot (McLaren Vale) Hay Shed Hill Wines (Margaret River) Heritage Estate (Granite Belt)

Howard Vineyard (Adelaide Hills) Ibis Wines (Orange) Idlewild (Hunter) Jarvis Estate (Margaret River) Jenke Vineyards (Barossa) Kotai Estate (Geographe) Lane's End Vineyard (Macedon Ranges) Leabrook Estate (Adelaide Hills) Longview Creek (Sunbury) Louee Wines (Mudgee) Mardia Wines (Barossa) Mount Avoca (Pyrenees) Mount Eyre Vineyards (Hunter) Old Loddon Wines (Bendigo) Paracombe Wines (Adelaide Hills) Passing Clouds (Bendigo) Pepper Tree Wines (Hunter) Peter Lehmann (Barossa) Plantagenet (Mount Barker) Polleters Vineyard (Pyrenees) Portree Vineyard (Macedon Ranges) Redgate (Margaret River) Rimfire Vineyards (Darling Downs) Ross Hill Wines (Orange) Settlement Wines (McLaren Vale) Sharpe Wines of Orange (Orange) SpringLane (Yarra Valley) St Leonards (Rutherglen) Steels Creek Estate (Yarra Valley) Swooping Magpie (Margaret River) Tahbilk (Nagambie Lakes) Tamborine Estate Wines (Queensland Coastal) Tamburlaine (Hunter) Truffle Hill Wines (Pemberton) Twelve Acres (Nagambie Lakes) Watershed Wines (Margaret River) Wild Dog Winery (Gippsland) Woodlands (Margaret River)

CABERNET SAUVIGNON

This variety is used in many regions throughout Australia, but as it is a mainstream variety it is beyond the scope of this book.

CARIGNAN

Synonyms: Carignan Noir, Bois Dur, Catalan, Carignane, Tinto Mazuella, Monestel, Roussillonen, Carinena

Maturity group: Very late ripening variety. Wine Type: Wines with super heavy body and with high acidity.

References: C, D, K, O

Carignan is a high producing red grape variety that is the staple of the production of vin ordinaire in the French Midi. It is widely grown for its yield but the wines are almost universally criticized by serious wine writers and critics. The area planted to this variety is declining rapidly in most areas where it is grown, chiefly in the warmer regions of France, Spain Italy as well as Australia – another case of growers responding to consumer demand for quality over quantity.

Carignan does have its supporters. Clos du Gravillas winery in the Minervois region of France is championing the variety in campaign dubbed Carignan Renaissance.

- Angoves (Riverland) Cascabel, (McLaren Vale) Happs (Margaret River) Kabminye Wines (Barossa) Kellermeister Wines (Barossa) Smallfry Wines (Barossa) Spinifex (Barossa)

CARINA

Maturity group: Early

References: K, O

This seedless red grape variety was bred by the CSIRO as a drying variety. It has found a use as a red wine variety noted for the production of deeply colored wine and is used to add colour in blends.

- Currans Family Wines (Murray Darling)

CARMENERE

Synonyms: Grand Vidure

Wine Type: Dry red wines

References: C, D, O

Carmenere is a red wine variety which was once popular in the Medoc District in Bordeaux. It fell from favour because of its susceptibility to the fungal disease. Like Petit Verdot, Carmenere is now quite uncommon in Bordeaux but it shows promise in the new world.

Nowadays this variety most commonly grown in Chile where many vineyards have found that the vines they thought were Merlot were in fact Carmenere. The vines were imported to Chile from Bordeaux in the nineteenth century and the mistake was not discovered until the 1990s.

As it turned out Carmenere has shown it is capable of producing high quality red wine, so many Chilean wineries have stuck with it. This story has echoes of the Sangiovese/Carnelian mix up in Western Australia, and of course the Albarino/Savagnin mix up unearthed in 2009.

Carmenere ripens earlier than Cabernet sauvignon and could be grown successfully in many Australian wine regions.

Wines made from Carmenere can have a strong green pepper flavour, especially if the grapes are picked early. Some of this flavour persists and adds to the wine if it is balanced out by other flavours.

- Amietta Vineyard (Geelong) Brown Brothers (King Valley) Macquarie Grove Vineyards (Western Plains) Olssens of Watervale (Clare Valley) Red Earth Estate (Western Plains) Ten Miles East (Adelaide Hills)

CARNELIAN

References: H, O

This variety originated in California as a cross of Cabernet sauvignon, this time with both Carignan and Grenache. It is intended as a hot climate alternative to Cabernet sauvignon and is often used to make wine for blending. It is grown in California, Texas and - believe it or not - in Hawaii.

Some years ago Howard Park Wines planted a vineyard in the Margaret River region with Carnelian, believing they were actually planting Sangiovese. It turned out a happy accident as they are now producing a powerful red wine, so even after the mistake was discovered the winemakers continued to use it. Other Western Australian vineyards also have Carnelian for a similar reason.

- Howard Park Wines (Margaret River) Jylland Vineyard (Central Western Australian Zone) McCuskers Vineyard (Perth Hills) Peos Estate (Manjimup) Western Range Wines (Perth Hills) Windshaker Ridge (Perth Hills)

CHAMBOURCIN

Synonyms: Joannes seyve 26.205

References: C, H, K, O

Chambourcin is a French Hybrid variety which has only been available since the 1960s. French hybrids are produced by crossing a variety of the European grape vine *Vitis vinifera* with an American vine species. The resulting vine has higher disease and pest resistance, but the wines produced often have an unusual or foxy flavour.

Since the 1960s French Hybrids have been systematically removed from much of France and hence are no longer of importance there. Chambourcin is reportedly grown in Vietnam.

Chambourcin is perhaps the most successful of the French Hybrids and is certainly the most widely used in Australia. The variety is especially resistant to fungal diseases so it is not surprising that it is most at home in the more humid regions of Northern NSW and Queensland.

Chambourcin wines are deeply coloured and fruity, but tend to finish short. Some producers use the variety for rose and sparkling reds. It is also used successfully for port style wine.

- Alderley Creek Wines Estate (Northern Rivers Zone) Allandale (Hunter) Allyn River Wines (Hunter) Apthorpe Estate (Hunter) Australian Old Vine Wines (Riverland) Bago Vineyards (Hastings River) Bawley Vale Estate (Shoalhaven Coast) Beelgara Estate (Riverina) Benwarin Wines (Hunter) Boatshed Vineyard (Hunter) Brierley Wines

(Queensland Zone) Calais Estate (Hunter) Cambewarra Estate (Shoalhaven Coast) Canolobas-Smith (Orange) Canungra Valley (Queensland Coastal) Capercaillie (Hunter) Cassegrain (Hastings River) Catspaw Farm (Granite Belt) Cedar Creek Estate (Queensland Coastal) Coolangatta Estate (Shoalhaven Coast) Cooper wines (Hunter) Crane Wines (South Burnett) Creeks Edge Wines (Mudgee) Crooked River Wines (Shoalhaven Coast) Cupitt's Winery (Shoalhaven Coast) D'Arenberg (McLaren Vale) Dayleswood Winery (Riverina) Divers Luck Wines (Northern Rivers Zone) Douglas Vale (Hastings River) Eumundi Winery (Queensland Coastal) Fairview Wines (Hunter) Fern Gully Winery (Shoalhaven Coast) Friday Creek Resort (Northern Rivers Zone) Frog Rock (Mudgee) Gowrie Mountain Estate (Darling Downs) Great Lakes (Northern Rivers Zone) Humphries Estate (Shoalhaven Coast) Inlam Estate (Northern Rivers Zone) Ivanhoe Wines (Hunter) Kingsley Grove (South Burnett) Long Point Vineyard (Hastings River) Lyrebird Ridge Organic Winery (South Coast Zone) Maleny Mountain Wines (Queensland Coastal) Maroochy Springs (Queensland Coastal) Mopoke Ridge Winery (Shoalhaven Coast) Mount Eyre Vineyards (Hunter) Mudgee Wines (Mudgee) Nightingale Wines (Hunter) Noosa Valley Winery (Queensland Coastal) Normanby Wines (Queensland Zone) Nowra Hill Vineyard (Shoalhaven Coast) Oak Works (Riverland) Oakvale (Hunter) O'Regan Creek Vineyard and Winery (Queensland Coastal) Pendarves Estate (Hunter) Petersons Glenesk Estate (Mudgee) Raleigh Wines (Northern Rivers Zone) Ridgeview Wines (Hunter) Roselea Estate (Shoalhaven Coast) Rothbury Ridge (Hunter) Rusty Fig Wines (South Coast Zone) Seven Mile Vineyard (Shoalhaven Coast) Sherwood Estate (Hastings River) Sirromet (Queensland Coastal) Southern Highland Wines (Southern Highlands) St Petrox (Hunter) Stonehurst Cedar Creek (Hunter) Tamburlaine (Hunter) Tilba Valley (South Coast Zone) Tinonnee Vineyard (Hunter) Twin Oaks (Queensland Coastal) Two Tails Wines (Northern Rivers Zone) Undercliff (Hunter) Villa d'Esta Vineyard (Northern Rivers Zone) Wright Family Wines (Hunter) Yarraman Estate (Hunter) Yarrawa Estate (Shoalhaven Coast)

CHARDONNAY

This variety is used in many regions throughout Australia, but as it is a mainstream variety it is beyond the scope of this book.

CHASSELAS

Synonyms: Fendant, Weisser Gutedel, Golden Chasselas, Chasselas Dore, Moster, Silberling, Marzemina Bianca, Dorin, Perlan, Tribianco Tedesco

Maturity group: Very early maturity, Wine Type: Light bodied white wines with low acidity.

References: C, D, H, K, O

This variety is used for table grapes and wine production although it is now much less popular as a wine grape. It is now rarely used for wine in France but continues to be used in Switzerland to produce Fendant dry white wine.

Until the 1990s it was grown by quite a few wineries in Western Victoria but most found it uneconomical to continue.

- Bullers Beverford (Swan Hill) Cathcart Ridge Estate (Grampians) Faranda Wines (Swan District) Villa d'Esta Vineyard (Northern Rivers Zone)

CHENIN BLANC

Synonyms: Steen, Pineau de la Loire, Pineau d'Anjou, Blanc d'Anjou, Gros Pineau de Vouvray, Franc-Blanc

Maturity group: This variety ripens in midseason, Wine Type: Full bodied wines with high acidity.

References: C, D, H, K, O

Chenin blanc is a versatile variety that can be used to make sweet, semi-sweet and dry white wines and also sparkling wine.

The famous Loire wines of Vouvray are made from Chenin blanc. The variety is also used for making wines under the Loire Valley appellations of Montlois, Savennieres and Saumur, to name just a few.

Chenin blanc is over cropped and undervalued almost everywhere else that it is used in the world, but some attention is at last being paid to its potential to produce quality wine when given due respect.

For many years the variety was the mainstay of the South African wine industry. This was largely due to the ability to produce large crops in warm conditions.

Most Australian plantings are in Western Australia, but Chenin is present in most Australian regions. It is grown extensively in Western Australia where it has formed the basis of the popular Houghton's White Burgundy. Along with Semillon and Sauvignon blanc and sometimes Chardonnay, Chenin blanc has been a component of the Western Australian Classic Dry White style.

It is probable that the WA propensity for using the variety is a result of importation of vines from South Africa in the 19th Century. Around Perth the variety has historically played a role similar to that it plays in South Africa, producing wines of no great character. The variety is reasonably popular in Margaret River where growers and winemakers have more of an eye for quality.

Elsewhere in Australia the variety plays a minor role in a number of regions, but it is fair to say the Australia has yet to produce a great Chenin blanc.

VARIETIES • Chenin blanc

- All Saints Estate (Rutherglen) Amberley Estate (Margaret River) Ambrook Wines (Swan Valley) Anderson Winery (Rutherglen) Bacchus Hill (Sunbury) Ballandeen Estate (Granite Belt) Bella Ridge Estate (Swan District) Berton Vineyards (Riverina) Blackboy Ridge Estate (Geographe) Blackwood Wines (Blackwood Valley) Blue Manna (Margaret River) Bremerton (Langhorne Creek) Broken River Vineyards (Goulburn Valley) Brookwood Estate (Margaret River) Brown Brothers (King Valley) Bullers Beverford (Swan Hill) Canungra Valley (Queensland Coastal) Cape Bouvard (Peel) Cape Grace Wines (Margaret River) Capel Vale (Geographe) Carabooda Estate (Swan District) Carilley Estate (Swan Valley) Carlaminda Estate (Geographe) Carosa (Perth Hills) Casley Mount Hutton Winery (Granite Belt) Chapman Valley Wines (Central Western Australian Zone) Chapman's Creek Vineyard (Margaret River) Charlies Estate Wines (Swan Valley) Cheriton (Swan District) Chidlows Well (Central Western Australian Zone) Chittering Valley Winery (Perth Hills) Cofield Wines (Rutherglen) Concotton Creek (Peel) Coriole (McLaren Vale) Cushendell (Southern Highlands) Darling Estate (King Valley) Dowie Doole (McLaren Vale) Drinkmoor Wines (Rutherglen) Evans and Tate (Margaret River) Fermoy Estate (Margaret River) Fish Tail Wines (Swan Valley) Flinders Bay (Margaret River) Flying Fish Cove (Margaret River) Fox River Wines (Mount Barker) Gabriel's Paddocks Vineyard (Hunter) Garbin Estate (Swan Valley) Goundrey (Mount Barker) Green Valley Vineyard (Margaret River) Halina Brook (Central Western Australian Zone) Happs (Margaret River) Harris Organic Wines (Swan Valley) Heafod Glen Winery (Swan Valley) Hotham Ridge Winery (Peel) Houghton (Swan Valley) Idylwild Wines (Geographe) Jane Brook Estate (Swan Valley) Jarrah Ridge Winery (Perth Hills) John Gehrig Wines (King Valley) Juniper Estate (Margaret River) Jylland Vineyard (Central Western Australian Zone) Kalleske Wines (Barossa) Kneedeep (Margaret River) Kominos Wines (Granite Belt) Kotai Estate (Geographe) Kyotmunga Estate (Perth Hills) Lamonts (Central Western Australian Zone) Lancaster Wines (Swan Valley) Lilac Hill Estate (Swan Valley) Little River Wines (Swan Valley) Longview Creek (Sunbury) Marri Wood Park (Margaret River) Middlebrook Estate (McLaren Vale) Moama Wines (Perricoota) Moondah Brook (Swan Valley) Murray Estate (North East Victoria) Myalup Wines (Geographe) Oakover Estate (Swan Valley) Olive Farm (Swan District) Paul Conti Wines (Perth) Peel Estate (Peel) Peter Lehmann (Barossa) Pikes (Clare Valley) Pinelli (Swan Valley) Plantagenet (Mount Barker) Pollocksford Vineyards (Geelong) Redgate (Margaret River) Riseborough Estate (Swan District) Riverbank Estate (Swan Valley) Riverina Estate Wines (Riverina) Romavilla (Roma) Rusden Wines (Barossa) Sandalford Wines (Swan Valley) Scarpatoni Estate (McLaren Vale) Settlers Ridge (Margaret River) Sevenhill Wines (Clare Valley) Sittella (Swan Valley) St Leonards (Rutherglen) Stakehill Estate (Peel) Strathkellar (Goulburn Valley) Sussanah Brook Wines (Swan District) Swan Valley Wines (Swan Valley) Swanbrook Estate Wines (Swan Valley) Swings & Roundabouts (Margaret River) Swooping Magpie (Margaret River) Talijancich (Swan Valley) Tassell Park Wines (Margaret River) Tatachilla (McLaren Vale) Temple Bruer (Langhorne Creek) The Lily Stirling Range (Great Southern) The Natural Wine Company (Swan Valley) Trappers Gully (Mount Barker) Tuart Ridge (Peel) Upper Reach Vineyard (Swan Valley) Valley Wines (Swan District) Vineyard 28 (Geographe) Vino Italia (Swan Valley) Voyager Estate (Margaret River) Wandering Brook Estate (Peel) Warraroong Estate (Hunter) Western Range Wines (Perth Hills) Westfield (Swan Valley) Windance Wines (Margaret River) Windows Margaret River

(Margaret River) Windy Creek Estate (Swan Valley) Woody Nook (Margaret River) Wordsworth Wines (Geographe) Yarran (Riverina)

CIENNA

Wine Type: Light wine with high acidity.

References: C, H, O

Cienna is a cross of Cabernet sauvignon and Sumoll, bred and developed by CSIRO for Australian conditions. One of its parents, Sumoll, from the Barcelona region, is noted for its adaptation to warm conditions. Cienna itself produces light bodied wines with strong colour and pronounced berry flavours.

- Brown Brothers (King Valley) Yalumba Wine Company (Barossa)

CINSAUT

Synonyms: Blue imperial, Black Prince, Oeillade, Ulliade, Ottavianello, Malaga, Cinsault, Plant d'Arles, Picardan Noir, Prunella, Espagne, Calabre, Cuviller, Poupe de Crabe, Black Malvoise

Maturity group: This variety ripens in midseason, Wine Type: Wines with super heavy body and with high acidity.

References: D, H, K, O

Cinsaut is high yielding 'workhorse' red variety, also commonly known as Blue Imperial. It was formerly used extensively for vin ordinaire in South of France and in other Mediterranean countries, but is now out of favour. The large number of synonyms is indicative of how widespread this variety used to be.

This variety, according to Australian wine writer Robin Bradley, is "used to add mediocrity to otherwise good wine". More accurately, it was used because it yields well.

It is most often used as blending material, probably in more wines than mention the fact on their labels. Cinsaut is also used for fortified wine production. Cinsaut makes quite good rose; Foggo Wines in McLaren Vale successfully use it this way

- Bullers Beverford (Swan Hill) Cascabel (McLaren Vale) Chambers Rosewood (Rutherglen) D'Arenberg (McLaren Vale) Donnybrook Estate (Geographe) Foggo Wines (McLaren Vale) Happs (Margaret River) Kabminye Wines (Barossa) Lake Moodemere (Rutherglen) Little River Wines (Swan Valley) Morris (Rutherglen) Massena (Barossa) Murray Street Vineyard (Barossa) Smallfry Wines (Barossa) Spinifex (Barossa) Vintara (Rutherglen)

CLAIRETTE

Synonyms: Blanquette, Clairette Blanc. There are several other varieties which have synonyms using the term 'clairette', for example Ugni blanc is known as Clairette Ronde.

Maturity group: Among the latest ripening varieties. Wine Type: Wines with moderate body and with low acidity.

References: C, D, H, K, O

This is another of those varieties from Southern France which yield prodigiously and generally produce fairly forgettable wines. While Clairette's role in contributing to the European Wine Lake are over it still has a part to play in blended wines, for example as one of the permitted varieties in the Chateauneuf-du-Pape appellation.

Clairette was once used in many vineyards, especially in the Hunter Valley, but it has all but disappeared. A notable exception is Honeytree Estate which produces a dry white which has gathered a cult following and always sells out soon after release.

- Honeytree Estate (Hunter) jb Wines (Barossa)

COLOMBARD

Synonyms: French Colombard

Maturity group: Late ripening variety, Wine Type: Very light bodied wines with high acidity.

References: C, D, H, K, O

This white wine variety is used extensively in the Charente region of France for base wines to be distilled for Cognac. But it has a wider role in South West France where it is often used for blended wines. It is also quite popular in California.

In Australia Colombard is ranked fifth in terms of volume of production, but almost all of the wine is sold as generic cask wine. The most notable exception perhaps is Primo Estate whose La Biondina Colombard is highly regarded. This further reinforces the point that many varieties with a reputation for producing large volumes of ordinary wine are not to be despised. With care and skill in the vineyard and winery these varieties may be capable of producing interesting wine with peach and nectarine aromas. Its ability to maintain high acid levels in warm conditions is also an asset.

At best Colombard can add acidity and peach and nectarine flavours to wines, but often the wines are bland and flat.

- Angoves Winery (Riverland) Australian Old Vine Wines (Riverland) Carn Estate (Murray Darling) Chateau Champsaur (Central Ranges Zone) De Bortoli (Riverina) Deakin Estate (Murray Darling) Gin Gin Wines (Queensland Coastal) Jaengenya Vineyard (Goulburn Valley) Norse Wines (Queensland coastal) Nyora Vineyard and Winery (Gippsland) Piromit Wines (Riverina) Primo Estate (Adelaide Plains) Red Tail (Northern Rivers Zone) Rimfire Vineyards (Darling Downs) Rodericks (South Burnett) Rossiters (Murray Darling) Salisbury Winery (Murray Darling) Sand Hills Vineyard (Central Ranges Zone) Steels Creek Estate (Yarra Valley) Stuart Range Estate (South Burnett) Sweet Water Hills Wines (Sunshine Coast) Trentham Estate (Murray Darling) Whiskey Gully Wines (Granite Belt)

CORTESE

Wine Type: Light bodied wines with moderate acidity.

References: C, D, H, O

This is a white grape variety from Piedmont and Lombardy North West Italy. It is the variety behind the famous Gavi wine.

Lost Valley Winery, in the Central Victorian Highlands has reputedly the only Cortese plantings outside the rare variety's Italian home. Lost Valley Cortese has received acclaim from wine writers and critics. It is worth seeking out.

- Lost Valley (Upper Goulburn)

CORVINA

Synonyms: Cruina, Corvina Veronese

Maturity group: Late ripening variety, Wine Type: Wines with moderate body and with moderate acidity.

References: C, D, O

This is a red wine variety from Veneto in North Eastern Italy. It is best known as the key variety in the Valpococella and Bardolino DOC's. The variety gives an almond flavour to the wines of which it is a component. It is also used in dried grape wine styles such as Amarone and Recioto.

The variety is rare in Australia, but Freeman Vineyards in the Hilltops Region has recently released a Rondinella Corvina blend made in the Amarone style.

- Freeman (Hilltops) Centennial Vineyards (Southern Highlands)

COUNOISE

Maturity group: Late ripening Wine Type: Very full bodied wines with high acidity.

References: C, D, H

Small areas are devoted to this red wine variety in the Rhone, Provence and Languedoc regions of France. It is prized for its ability to add peppery, spicy fruit flavours to blends such as Chateauneuf-du-Pape.

Battely are planning to establish the variety and hope to produce wine in the next few years

- Battely Wines in Beechworth

CROUCHEN

Synonyms: Clare Riesling, Cape Riesling, Messange Blanc, Navarre Blanc, Sable blanc

Maturity group: This variety ripens in midseason.

References: C, K, O

This French white variety has been largely abandoned around the world. It makes wine which is fairly neutral and has been replaced by more appealing varieties. Until the 1970s Crouchen was relatively popular in many Australian regions under the name of Clare Riesling. These days one of the last strongholds of the variety is Brown Brothers Winery which makes a semi sweet Crouchen Riesling blend which they market as a Moselle style.

- Andrew Peace Wines (Swan Hill) Botobolar (Mudgee) Brown Brothers (King Valley) Jillian Wines (Grampians) Reads (King Valley) Romavilla (Roma) Taylors (Clare Valley)

CYGNE BLANC

References: C

This white wine variety arose from a seedling discovered in 1989 near a Cabernet sauvignon vineyard in the Swan Valley. Thus it qualifies as an Australian variety, along with those deliberately bred by the CSIRO and the Shalistin and Malian varieties which arose as sports of existing vines.

Mann in the Swan Valley uses Cygne blanc in a blend with Cabernet sauvignon to make a sparkling wine. Port Robe Estate has a large planting in a vineyard in the Mount Benson region where they intend to make a still white wine from it.

The story of Cygne blanc shows how a new variety can arise. Some one notices an individual vine with a special characteristic, and then devotes some time and effort to propagate from it, grow some grapes, make some wine and evaluate and promote the final product. It requires luck, vision and lots of persistence.

- Port Robe Estate (Mount Benson), Mann (Swan Valley)

DOLCETTO

Synonyms: Dolsin, Charbono, Ormaesco

Maturity group: Early maturing, Group 3, Wine Type: Very full bodied wines with low acidity.

References: D, H, K, O

Dolcetto is an Italian red wine grape variety which is popular in the cooler northern regions of Italy, particularly in Piedmont region.

Its early ripening makes it a feasible choice in areas where Barbera and Nebbiolo would struggle to ripen.

Although the name implies sweetness, Dolcetto wines are usually dry. The variety is noted for producing relatively low acid wines. The strong dark colour of the grape is reflected in the deep colour of the wine. Low to moderate acidity and tannins mean that these wines are best consumed fairly young, and they are rarely great wines.

The variety hasn't taken off in Australia like some of the other red Italian varieties, such as the increasingly popular Sangiovese. However it seems that Australia is the only country outside Italy that has significant plantings of the variety.

Brown Brothers blends Dolcetto with Shiraz, or Syrah as it says on the label to make a popular sweetish red. Unfortunately this wine has perpetuated the idea that Dolcetto wines are sweet.

- Amadio (Adelaide Hills) Bests (Grampians) Box Stallion (Mornington Peninsula) Brown Brothers (King Valley) Catherine Vale Vineyard (Hunter) Ciavarella (King Valley) Crittenden at Dromana (Mornington Peninsula) Dromana Estate (Mornington

Peninsula) Gracebrook Vineyards (King Valley) Heartland Wines (Limestone Coast) La Cantina King Valley (King Valley) Massena Wines (Barossa) Mount Franklin Estate (Macedon Ranges) Parish Hill Wines (Adelaide Hills) Rojo Wines (Port Phillip Zone) Snobs Creek Wines (Upper Goulburn) Symphonia (King Valley) Turkey Flat Vineyards (Barossa) Vale Creek Wines (Central Ranges Zone) Vintara (Rutherglen) Wanted Man (Heathcote) Warburn Estate (Riverina) Warrenmang Vineyard (Pyrenees) Yacca Paddock Vineyards (Adelaide Hills) Zonte's Footstep (Langhorne Creek)

DORADILLO

Maturity group: Among the latest ripening varieties, Wine Type: White wine for distillation

References: H, K, O

This is a Spanish variety but it is believed extinct in its native country. Doradillo's ability to produce very high yields in warm irrigated regions contributed to its popularity for cask wines and for its continuing role as material for distillation.

- Angoves Winery (Riverland) Elliot Rock Winery (Mudgee)

DURIF

Synonyms: Petite Sirah, Pinot d'Hermitage, Syrah Fourche

Maturity group: This variety ripens just before midseason. Wine Type: Wines with super heavy body and with moderate acidity.

References: C, D, H, K, O

Like many grape varieties Durif is making a comeback after being overlooked in the rush by vignerons to get with the strength and concentrate on mainstream varieties. Some great red wines are now being made in many Australian wine regions using Durif with the label proudly identifying the variety.

Durif is not highly regarded in its native Rhone Valley. It is named after its breeder who established the variety in the late nineteenth century. The variety is planted in California under the name of Petit Sirah where there has been some confusion about the variety. Many Californian Petite Sirah vineyards have proved to have mixtures of Durif and Pelorisin.

Durif was introduced to Australia in 1908 in the Rutherglen district after encouragement by Francois de Castella, Victoria's influential pioneering viticulturalist.

VARIETIES • Durif

Although Durif has been traditionally associated with Rutherglen and warmer climates the Vale Vineyard on the Mornington Peninsula report success using it in a cooler region. This is in accord with Gladstones placing the variety in Maturity Group 3, and thus ripening earlier than say Shiraz, Cabernet sauvignon and Merlot.

Wines made from this variety tend to have high levels of tannins and so are suitable for extended cellaring to allow the flavours to come together. Modern winemakers of course aim to produce wines which will be drinkable much earlier so this may not be such a factor as it was in the past.

Dan Crane, winemaker at St Leonards and All Saints Estate believes that the key to handling the tannins in Durif is to use extra long maceration on the skins to produce softer tannins. However if you are looking for a subtle and understated red you should look elsewhere. The alcohol level in the wines is often in the 14-15% range, with flavour to match. Enjoy them with rich stews and game dishes.

- 919 Wines (Riverland) All Saints Estate (Rutherglen) Anderson Winery (Rutherglen) Battely Wines (Beechworth) Beelgara Estate (Riverina) Bellarine Estate (Geelong) Boyntons Feathertop (Alpine Valleys) Brown Brothers (King Valley) Brumby Wines (Swan Hill) Bullers Calliope (Rutherglen) Calais Estate (Hunter) Campbells Wines (Rutherglen) Cape Horn Vineyard (Goulburn Valley) Ciavarella (King Valley) Cofield Wines (Rutherglen) Connor Park (Bendigo) Currans Family Wines (Murray Darling) Date Brothers (Swan Hill) Dos Rios (Swan Hill) Drinkmoor Wines (Rutherglen) Eumundi Winery (Queensland Coastal) Gapsted (Alpine Valleys) Gehrig Estate (Rutherglen) Golden Grove Estate (Granite Belt) Heritage Estate (Granite Belt) Jaengenya Vineyard (Goulburn Valley) John Gehrig Wines (King Valley) Judds Warby Range Estate (Glenrowan) Kingston Estate (Riverland) Lake Moodemere (Rutherglen) Maleny Mountain Wines (Queensland Coastal) Massena Wines (Barossa) Melange Wines (Riverina) Michael Unwin Wines (Grampians) Miranda (Riverina) Miranda Wines, Griffith (Riverina) Morris (Rutherglen) Morrisons of Glenrowan (Glenrowan) Mount Pilot Estate (North East Victoria) Mount Prior (Rutherglen) Mudgee Wines (Mudgee) Myattsfield Vineyard and Winery (Perth Hills) New Glory (Goulburn Valley) Normanby Wines (Queensland Zone) Nugan Estate (King Valley) Oak Works (Riverland) Paradine Estate (Queensland Zone) Patrice Winery (North East Victoria) Petersons Glenesk Estate (Mudgee) Piako Vineyards (Murray Darling) Piromit Wines (Riverina) Pyren Vineyard (Pyrenees) Reedy Creek (Northern Slopes Zone) Renewan (Swan Hill) Riverina Estate Wines (Riverina) Rose Hill Estate Wines (King Valley) Rothbury Ridge (Hunter) Rusticana (Langhorne Creek) Rutherglen Estates (Rutherglen) Sam Miranda Wines (King Valley) Scion Vineyard (Rutherglen) St Petrox (Hunter) Stanton and Killeen Wines (Rutherglen) Steinborner Family Vineyards (Barossa) Taminick Cellars (Glenrowan) Tawonga Vineyard (Alpine Valleys) Tinonnee Vineyard (Hunter) Toolleen Vineyard (Heathcote) Trahna Rutherglen Wines (Rutherglen) Vale Vineyard (Mornington Peninsula) Valhalla Wines (Rutherglen) Vintara (Rutherglen) Warrabilla Wines (Rutherglen) Watchbox Wines (Rutherglen) Wedgetail Ridge Estate (Darling Downs) Westend Estate (Riverina) Wirruna Estate

(North East Victoria) Witches Falls Winery (Granite Belt) Yacca Paddock Vineyards (Adelaide Hills)

FIANO

Wine Type: Wines with moderate body and with moderate acidity.

References: C, D, H, O

Fiano is a white wine grape from Campania in Southern Italy. It is the basis for the wine Fiano di Avellino, noted for its honey and spice aroma. Although the variety is quite rare Fiano is being pioneered in Australia as it has the capacity to produce well structured wine in warm to hot conditions.

At the 2009 Australian Alternative Varieties Wine Show the organisers created separate class for this variety, indicative of their faith in its future in Australia.

- Banrock Station (Riverland) Beach Road (Langhorne Creek) Chalmers (Murray Darling) Coriole (McLaren Vale) Fox Gordon (Barossa) Karanto Vineyards (Langhorne Creek) Olivers Taranga (McLaren Vale) Parish Hill Wines (Adelaide Hills) Rutherglen Estates (Rutherglen) Sutton Grange Winery (Bendigo) Witches Falls Winery (Granite Belt)

FLORA

References: H, K, O

This aromatic white wine variety is the result of a crossing of Gewurztraminer and Semillon. Brown Brothers use it in a blend with Orange Muscat.

Brown Brothers (King Valley)

FRAGOLA

Synonyms: Isabella

The name of this red wine grape means 'strawberry' in Italian. The wines it produces have an unmistakable strawberry bouquet. Fragola is popular with amateur Italian winemakers in Australia.

- Bacchus Hill Winery (Sunbury) Douglas Vale (Hastings River) Folino Estate (Alpine Valleys) Michelini (Alpine Valleys) Pasut Family Wines (Murray Darling)

The codes for the References (C, D, H, K, O) are listed on page 9 of this book

FURMINT

Synonyms: Sipon, Moslavac Bijeli

Maturity group: This variety ripens just before midseason. Wine Type: Full bodied wines with high acidity.

References: C, D, H, K, O

This variety is the basis for the Tokaji, the fine desert wine from Hungary. It is characterized by finesse, complex flavours and ageing potential. Furmint is used in Hungary and in neighboring countries for dry whites and for botrytised styles.

- Briery Estate (Perth Hills), Eling Forest Estate (Southern Highlands) Happs (Margaret River)

GAMAY

Synonyms: Gamay Noir, Petit Gamay, Bourgiunon Noir, Gamay Rond

Maturity group: Early maturing, Group 3, Wine Type: Wines with moderate body and with moderate acidity.

References: C, D, H, K, O

Gamay is best known for its role in Beaujolais, but it also plays a role in producing spicy reds and roses in the lower Loire Valley. In Burgundy Gamay plays second fiddle to Pinot Noir as a component in lesser wines. In Switzerland it is plays a similar role where it is blended with Pinot Noir to produce a wine known as Dole. There are quite a few plantings of Gamay in Australia across a broad range of climate, from Southern Tasmania, the Mornington Peninsula, North Eastern Victoria, Hunter Valley and the Granite Belt in Queensland.

Gamay struggles against the perception that it makes only light bodied wine, but the best wines from this variety are full flavoured.

- Bass Phillip (Gippsland) Brave Goose Winery (Goulburn Valley) Cofield Wines (Rutherglen) Cushendell (Southern Highlands) Darling Estate (King Valley) Elan Vineyard (Mornington Peninsula) Eldridge Estate (Mornington Peninsula) Evans Family Wines (Hunter) Fernbrook Estate (Porongurup) Gowrie Mountain Estate (Darling Downs) Grandview Vineyard (Southern Tasmania) Happs (Margaret River) John Gehrig Wines (King Valley) Lawson Hill (Mudgee) Little Bridge (Canberra) Marybrook Vineyards & Winery (Margaret River) Mount Burrumboot Estate (Heathcote) Pennyweight Winery (Beechworth) Pfeiffer Wines (Rutherglen) Red Earth Estate (Western Plains) Roundstone Winery (Yarra Valley) Sailors Falls Winery

(Macedon Ranges) Scarpatoni Estate (McLaren Vale) Sorrenberg (Beechworth) Sutherlands Creek Vineyard (Geelong) Wilmot Hills Vineyard (Northern Tasmania)

GARGANEGA

Synonyms: Gargana, Lizzana, Ostesona

Wine Type: Light bodied wines with low acidity.

References: C, D, H, O

This is the white grape variety used to produce Soave and other Northern Italian wines. It is highly vigorous prone to overproduction and thus to fairly bland wines. It can also produce sweet *Recioto* style wines which are made from raisined grapes.

Domain Day (Barossa) Romavilla (Roma)

GEWURZTRAMINER

Synonyms: Traminer, Rotclevner, Savagnin Rose, Fermin Rouge, Fromenteau Rouge, Ranfolzia, Kleinwiner, Drumin, Heida, Mala Dinka, Pinat Cervana, Liwora

Maturity group: Early maturing. Wine Type: Very full bodied aromatic wines with low acidity.

References: C, D, H, O

Gewurztraminer is best known for its wines from Alsace but it is widely planted elsewhere in France as well as in Germany and Central and Eastern Europe.

In Australia it is also known as Traminer, especially in the warmer regions where it is most commonly used as a blending companion for Riesling.

In cooler Australian wine regions Gewurztraminer produces highly aromatic wines. The strong aroma of Gewurztraminer leads many to assume that they are sweet wines. Some are in fact dry, and in any case these are serious wines that deserve more attention than they get.

- 5 Corners Wines (Mudgee) 572 Richmond Road (Southern Tasmania) Angoves Winery (Riverland) Ashton Hills (Adelaide Hills) Audrey Wilkinson (Hunter) Bay of Fires (Northern Tasmania) Beelgara Estate (Riverina) Belgravia Vineyards (Orange) Bianchet (Yarra Valley) Birthday Villa Vineyard (Macedon Ranges) BK Wines (Adelaide Hills) Blackets (Adelaide Hills) Borrodell on the Mount (Orange) Bowmans Run (Beechworth) Bracken Hill (Southern Tasmania) Brammar Estate (Yarra Valley) Bream Creek Vineyard (Southern Tasmania) Briar Ridge Vineyard (Hunter) Bungawarra (Granite Belt) Burke and Wills Winery

36 VARIETIES • Gewurztraminer

(Heathcote) Capercaillie (Hunter) Cargo Road Wines (Orange) Cassegrain (Hastings River) Chatsfield (Mount Barker) Conte Estate Wines (McLaren Vale) Contessa Estate (McLaren Vale) Craigow (Southern Tasmania) Cross Rivulet (Southern Tasmania) Delatite Winery (Upper Goulburn) Elliot Rocke Estate (Mudgee) Elsewhere Vineyard (Southern Tasmania) Ernest Hill Wines (Hunter) Felsberg Winery (Granite Belt) Fernfield Wines (Eden Valley) Frogmore Creek Vineyard (Northern Tasmania) Gin Gin Wines (Queensland Coastal) Glen Erin Vineyard Resort (Macedon Ranges) Gnadenfrei Estate (Barossa) Golden Grape Estate (Hunter) Grandview Vineyard (Southern Tasmania) Grevillea Estate (South Coast Zone) Hainault (Perth Hills) Henderson Hardie (King Valley) Henschke (Eden Valley) Heritage Farm (Goulburn Valley) Hidden Creek (Granite Belt) Hillside Estate (Hunter) Hood Wines (Southern Tasmania) Hunting Lodge Estate (South Burnett) Huntleigh Vineyards (Heathcote) Iron Pot Bay Wines (Northern Tasmania) Ivanhoe Wines (Hunter) Jamsheed (Yarra Valley) Jindalee Estate (Geelong) Josef Chromy Wines (Northern Tasmania) Kangderaar Vineyard (Bendigo) Kellermeister Wines (Barossa) Kellybrook (Yarra Valley) Kevin Sobels Wines (Hunter) Kirkham Estate (Sydney Basin) Kladis Estate (Shoalhaven Coast) Knappstein Wines (Clare Valley) Kreglinger Estate (Mount Benson) Lambert Vineyards (Canberra) Lancefield Winery (Macedon Ranges) Lawson Hill (Mudgee) Leabrook Estate (Adelaide Hills) Lillydale Estate (Yarra Valley) Lillypilly Estate (Riverina) Little Wine Company (Hunter) Logan Wines (Mudgee) Long Point Vineyard (Hastings River) Lyre Bird Hill (Gippsland) Manton's Creek Vineyard (Mornington Peninsula) Midhill Vineyard (Macedon Ranges) Milton Vineyard (Southern Tasmania) Miramar (Mudgee) Moorebank Vineyard (Hunter) Moorilla Estate (Southern Tasmania) Mount Macedon Winery (Macedon Ranges) Mudgee Wines (Mudgee) No Regrets (Southern Tasmania) Pepper Tree Wines (Hunter) Petersons Glenesk Estate (Mudgee) Pewsy Vale (Eden Valley) Pipers Brook Vineyard (Northern Tasmania) Pirie Estate (Northern Tasmania) Platt's (Rutherglen) Plunkett Fowles (Strathbogie Ranges) Point Leo Road Vineyard (Mornington Peninsula) Ridgeview Wines (Hunter) Robert Stein (Mudgee) Rosevears Estate (Northern Tasmania) Rymill Coonawarra (Coonawarra) Sailors Falls Winery (Macedon Ranges) Seldom Seen (Mudgee) Sevenhill Wines (Clare Valley) Shepherds Run (Canberra) Skillogalee (Clare Valley) Sorby Adams (Eden Valley) Souters Vineyard (Alpine Valleys) Southern Grand Estate (Hunter) Southern Highland Wines (Southern Highlands) Spring Vale Wines (Southern Tasmania) Stonehaven (Padthaway) Straws Lane (Macedon Ranges) Sutherland Estate (Yarra Valley) Taliondal (Hunter) Tamar Ridge (Northern Tasmania) Tanglewood Downs (Mornington Peninsula) Taylors (Clare Valley) The Minya Winery (Geelong) The Silos Estate (Shoalhaven Coast) The Wanderer (Yarra Valley) Tilba Valley (South Coast Zone) TK Wines (Adelaide Hills) Tomich Hill (Adelaide Hills) Toms Cap (Gippsland) Toppers Mountain (New England) Transylvania Winery (Southern New South Wales Zone) Two Tails Wines (Northern Rivers Zone) Vicarys (Sydney Basin) Vincognita (McLaren Vale) Virage (Margaret River) Walsh Family Winery (Perth Hills) Warrenmang Vineyard (Pyrenees) Westend Estate (Riverina) Whyworry Wines (New England) Wili-Wilia Winery (Macedon Ranges) Wilson Vineyard (Clare Valley) Windy Ridge Vineyard (Gippsland) Winooka Park (Central Ranges Zone) Woodonga Hill (Hilltops) Wroxton Wines (Eden Valley) Yarraman Estate (Hunter) Yass Valley Wines (Canberra)

•

GOUAIS BLANC

References: C, H, K, O

This variety seems to all but disappeared from it Europe, but its importance lies beyond its current use. It is the ancestor of many modern varieties including Chardonnay.

Chambers make a dry white wine from their small plantation of Gouais.

- Chambers Rosewood (Rutherglen)

GRACIANO

Synonyms: Morrastel, Perpignanou Bois Dur, Xeres, Courouillade, Couthurier, Tinta Muida

Maturity group: Very late ripening variety, Wine Type: Very full bodied wines with low acidity.

References: C, D, H, K, O

Graciano is red grape variety that is making a comeback in Rioja where it plays second fiddle to Tempranillo. It is also grown in Southern France where it carries the name Morrastel. In Spain the term Morrastel refers to the variety Monastrell, or Mourvedre. Whatever, the wine is strongly aromatic and can make interesting blending material as well as a straight varietal.

In Australia Graciano is used to make varietal wines and in blends with Tempranillo. It is probably better suited to the latter role.

- Artwine (Clare Valley) Back Pocket (Granite Belt) Brown Brothers (King Valley) Cascabel (McLaren Vale) Ciavarella (King Valley) Donnybrook Estate (Geographe) Epsilon (Barossa) Happs (Margaret River) Lillian (Pemberton) Mazza (Geographe) Moss Brothers (Margaret River) Mount Majura (Canberra) Rimfire Vineyards (Darling Downs) Ross Estate Wines (Barossa) Rudderless Wines (McLaren Vale) Swings & Roundabouts (Margaret River) Talijancich (Swan Valley) The Grove Vineyard (Margaret River) Tscharke (Barossa) Vinifera Wines (Mudgee) Zonte's Footstep (Langhorne Creek)

GRECANICO

Synonyms: Grecanico Dorato

References: C O

Grecanico is white wine variety from Sicily. It has a grassy aromatics in a way similar to Sauvignon blanc, but presumably is more suitable for warmer climates.

- Politini (King Valley)

GRECO BIANCO

References: C, O

Greco is a southern Italian white variety, presumably of Greek origins. It is used to make the dry Greco di Tufo in Campania Italy, and the sweet Greco di Bianco of Calabria.

Greco is being pioneered in Australia along with other Southern Italian varieties in the belief that Australian climates are more akin to the south, rather than the north of Italy.

- Beach Road (Langhorne Creek) Chalmers (Murray Darling) Karanto (Langhorne Creek)

GRENACHE

Synonyms: Garnacha, Granaccia, Lladoner, Tinto, Tinta, Carignan rosso, Carignane rousse, Tinto arogonese, Uva di Spagna, Sans Pariel, Tinto Menudo, Tinta MencidaTintilo de Rota, Rousillon, Rousillon Tinto Rouvaillard, Cannonau, Alicante, Rivesaltes, Redondal (and probably lots more...)

Maturity group: Very late ripening variety, Wine Type: Wines with super heavy body and with high acidity.

References: C D, H, K, O

While Grenache has a long history in Australia, it is only over the past decade or so that the variety has received the recognition that it deserves. While it is regarded as a classic variety by many writers it is an 'alternative varietal' in Australia, not least because few wine drinkers know much about it.

This grape variety is widely planted in South Australia, particularly in the Barossa Valley and McLaren Vale wine regions. It is a versatile variety which can be used as a straight varietal wine, it makes very good rose and is used as blending material, particularly with Shiraz and Mourvedre. In fact the so called GSM blends are becoming a signature Barossa/McLaren Vale style, challenging the dominance of straight Shiraz.

There is some contention among wine writers as to whether this variety should be regarded as French, or as a Spanish variety. In Spain it is known as

Garnacha, and is grown extensively throughout the North and East of the country.

In France this variety is grown in the Southern Rhone region as well as in Roussillon. In the Rhone it is a key ingredient of the famous Chateauneuf-du-Pape wines. In fact Grenache is a principal variety in all of the major Appellations in the Southern Rhone.

The ability of this variety to produce high yields when given plenty of irrigation in warm climates made it the most popular Australian variety until the 1960s. It was the basis of the so called port wines, as well as a component of many dry red wines, which were called claret in those days.

But only rarely did the word Grenache appear on a wine label. Shiraz and Cabernet Sauvignon replaced Grenache in the vineyards and the variety seemed destined for obscurity in Australia.

In the 1980s, this variety was subject of a vine-pull scheme. Fortunately a few growers and winemakers recognised the virtues of Grenache and it now is nudging along nicely in the wine fashion stakes. A few remaining paddocks of old vines have suddenly been elevated to the status of viticultural gold mines.

As young wines Grenaches are firm and sometimes rough textured. I suspect that getting the tannin balance right is a major factor in making a good Grenache. The colour often does not adequately prepare you for the depth of flavour.

While the lighter styles are versatile as accompaniments to food, bigger styles demand a meaty dish to really show their stuff. GSM wines are firm, well flavoured and coloured. They go well with barbeques, but are certainly not out of place with the best cuts of meat.

- Adelina Wines (Clare Valley) Amadio (Adelaide Hills) Amarillo Wines (Peel) Ambrook Wines (Swan Valley) Andrew Peace Wines (Swan Hill) Angas Vineyards (Langhorne Creek) Arakoon (McLaren Vale) Arimia Margaret River (Margaret River) Artwine (Clare Valley) Australian Domaine Wines (Clare Valley) B3 Wines (Barossa) Ballast Stone Estate (Currency Creek) Balthazar (Barossa) Barossa Valley Estate (Barossa) Barristers Block (Adelaide Hills) Basedow (Barossa) Bell River Estate (Central Ranges Zone) Bella Ridge Estate (Swan District) Bent Creek Vineyards (McLaren Vale) Bethany (Barossa) Biscay Wines (Barossa) Blackbilly (McLaren Vale) Blown Away (McLaren Vale) Boireann (Granite Belt) Briery Estate (Perth Hills) Brini Estate (McLaren Vale) Broken River Vineyards (Goulburn Valley) Brookhampton Estate (Geographe) Bullers Beverford (Swan Hill) Burge Family Winemakers (Barossa) Byramgou Park (Geographe) Callipari Wine (Murray Darling) Canonbah Bridge (Western Plains) Cape Barren Wines (McLaren Vale) Cape Bouvard (Peel) Cape Mentelle (Margaret River) Carilley Estate (Swan Valley) Carpenteri Vineyards (Swan Hill) Cascabel (McLaren Vale) Chain of Ponds (Adelaide Hills) Chaperon Wines (Bendigo) Charles Melton (Barossa) Chateau Dorrien (Barossa) Chittering Valley Winery (Perth Hills) Cirillo (Barossa) Clancy Fuller (Barossa) Clarence Hill (McLaren

40 VARIETIES • Grenache

Vale) Clarendon Hills (McLaren Vale) Classic McLaren Wines (McLaren Vale) Claymore Wines (Clare Valley) Cobbitty Wines (South Coast Zone) Colonial Estate (Barossa) Conte Estate Wines (McLaren Vale) Contessa Estate (McLaren Vale) Coriole (McLaren Vale) Craneford (Barossa) Currans Family Wines (Murray Darling) Currency Creek (Currency Creek) D'Arenberg (McLaren Vale) Darlington Estate (Perth Hills) De Lisio Wines (McLaren Vale) Deisen (Barossa) Di Fabio Estate (McLaren Vale) Diggers Bluff (Barossa) Diloreto Wines (Adelaide Plains) Doctors Nose Wines (New England) DogRidge (McLaren Vale) Dogrock Winery (Pyrenees) Domain Barossa (Barossa) Dominic Versace Wines (Adelaide Plains) Donnybrook Estate (Geographe) Eden Road Wines (Eden Valley) Eperosa (Barossa) Eyre Creek (Clare Valley) Faranda Wines (Swan District) Fireblock (Clare Valley) Five Geese Hillgrove Wines (McLaren Vale) Flat View Vineyard (Clare Valley) Foggo Wines (McLaren Vale) Fonthill Wine (McLaren Vale) Fox Creek Wines (McLaren Vale) Franand Wines (Swan District) FUSE (Clare Valley) Gapsted (Alpine Valleys) Gawler River Grove (Adelaide Plains) Geddes Wines (McLaren Vale) Gemtree Vineyards (McLaren Vale) Geoff Hardy (McLaren Vale) Geoff Merrill (McLaren Vale) Gibson Barossavale (Barossa) Gilligan (McLaren Vale) Gin Gin Wines (Queensland Coastal) Glaetzer Wines (Barossa) Glaymond Wines (Barossa) Gnadenfrei Estate (Barossa) Golden Ball (Beechworth) Gomersal Wines (Barossa) Gomersal Wines (Barossa) Grancari Estate (McLaren Vale) Grant Burge (Barossa) Greenock Creek Wines (Barossa) Halina Brook (Central Western Australian Zone) Hamiltons Ewell Vineyards (Barossa) Hanging Rock Winery (Macedon Ranges) Harman's Ridge Estate (Margaret River) Haselgrove (McLaren Vale) Hawkers Gate (McLaren Vale) Heathcote Estate (Heathcote) Henry Holmes Wines (Barossa) Henschke (Eden Valley) Hently Farm Wines (Barossa) Hewitson (Barossa) Hickinbotham (Mornington Peninsula) Hobbs of Barossa Ranges (Barossa) Hutton Vale (Eden Valley) Izway Wines (Barossa) Jamabro Wines (Barossa) Jeanneret Wines (Clare Valley) Jenke Vineyards (Barossa) Jimbour Wines (Queensland Zone) John Duval Wines (Barossa) Jupiter Creek Winery (Adelaide Hills) Kabminye Wines (Barossa) Kaesler (Barossa) Kalleske Wines (Barossa) Kangarilla Road (McLaren Vale) Kay Bros Amery (McLaren Vale) Kellermeister Wines (Barossa) Kilikanoon (Clare Valley) Kirrihill (Adelaide Hills) Kirrihill Estates (Clare Valley) Kitty Crawford Estate (New England) Kladis Estate (Shoalhaven Coast) Koltz (McLaren Vale) Kyotmunga Estate (Perth Hills) La Curio (McLaren Vale) Lake Breeze (Langhorne Creek) Landhaus Estate (Barossa) Langmeil (Barossa) Lanzthomson Wines (Barossa) Laughing Jack (Barossa) Linda Domas Wines (McLaren Vale) Linfield Road Wines (Barossa) Little Bridge (Canberra) Lou Miranda Estate (Barossa) Macaw Creek Wine (Mount Lofty Ranges Zone) Magpie Estate (Barossa) Mansfield Wines (Mudgee) Marienberg (McLaren Vale) Mary Byrnes Wines (Granite Belt) Marybrook Vineyards & Winery (Margaret River) mas serrat (Yarra Valley) Massena Wines (Barossa) Maverick Wines (Barossa) Mawson Ridge (Adelaide Hills) Maxwell Wines (McLaren Vale) McHenry Hohnen (Margaret River) McLaren Ridge Estate (McLaren Vale) McLaren Wines (McLaren Vale) McWilliams (Riverina) Middlebrook Estate (McLaren Vale) Mitchell (Clare Valley) Mitchelton (Nagambie Lakes) Moppa Wilton Vineyards (Barossa) Morrisons Riverview Winery (Perricoota) Moss Brothers (Margaret River) Mount Appallan Vineyards (South Burnett) Mount Trafford (Southern Fleurieu) Mt Billy (Southern Fleurieu) Murdock (Barossa) Murray Street Vineyard (Barossa) Neagles Rock Vineyards (Clare Valley) Noon Winery (McLaren Vale) Normanby Wines (Queensland Zone) Old Plains (Adelaide Plains) Old Station (Clare Valley) Olive Farm (Swan District) Oliverhill

(McLaren Vale) Olivers Taranga (McLaren Vale) Organic Vignerons Australia (Riverland) Orlando (Barossa) Parkerville Ponds Vineyard (Perth Hills) Parri Estate (Southern Fleurieu) Patterson Lakes Estate (Port Phillip Zone) Paul Conti Wines (Perth) Paxton (McLaren Vale) Pende Valde (McLaren Vale) Penny's Hill (McLaren Vale) Pepperilly Estate Wines (Geographe) Pertaringa (McLaren Vale) Peter Lehmann (Barossa) Phoenix Estate (Clare Valley) Pikes (Clare Valley) Pirramimma (McLaren Vale) Plantagenet (Mount Barker) Possums Vineyard (McLaren Vale) RBJ (Barossa) Red Earth Estate (Western Plains) Redbox Perricoota (Perricoota) Redheads Studio (McLaren Vale) Reilly's Wines (Clare Valley) Remarkable View Winery (Southern Flinders Region) Richard Hamilton Wines (McLaren Vale) Riseborough Estate (Swan District) Riverbank Estate (Swan Valley) Rockford (Barossa) Roehr (Barossa) Roennfeldt Wines (Barossa) Rosenvale Wines (Barossa) Ross Estate Wines (Barossa) Rudderless Wines (McLaren Vale) Rusden Wines (Barossa) Rutherglen Estates (Rutherglen) Saltram (Barossa) Samuels Gorge (McLaren Vale) SC Pannell (McLaren Vale) Scarpatoni Estate (McLaren Vale) Schild Estate Wines (Barossa) Schiller Vineyards (Barossa) Schulz Vignerons (Barossa) Schwarz Wine Company (Barossa) Scion Vineyard (Rutherglen) Scorpiiion (Barossa) Serafino Wines (McLaren Vale) Seraph's Crossing (Clare) Sevenhill Wines (Clare Valley) Shingleback (McLaren Vale) Sieber Road Wines (Barossa) Simon Hackett (McLaren Vale) Skillogalee (Clare Valley) Smallfry Wines (Barossa) Sons of Eden (Barossa) Soul Growers (Barossa) Spinifex (Barossa) Spook Hill Wines (Riverland) Springs Hill Vineyard (Fleurieu Zone) St Annes Vineyards (Perricoota) St Hallett (Barossa) Stakehill Estate (Peel) Stonewell Vineyards (Barossa) Strathkellar (Goulburn Valley) Sutherlands Creek Vineyard (Geelong) Swan Valley Wines (Swan Valley) Swings & Roundabouts (Margaret River) Tahbilk (Nagambie Lakes) Tait Wines (Barossa) Talijancich (Swan Valley) Talunga (Adelaide Hills) Tatachilla (McLaren Vale) Te-Aro (Barossa) Temple Bruer (Langhorne Creek) Tenafeate Creek Wines (Adelaide Plains) Teusner (Barossa) The Gap (Grampians) The Grapes of Ross (Barossa) The Islander Estate Vineyards (Kangaroo Island) The Lily Stirling Range (Great Southern) The Minya Winery (Geelong) The Old Faithful Estate (McLaren Vale) The Ritual (Peel) Tim Adams (Clare Valley) Tim Smith Wines (Barossa) Tin Shed Wines (Eden Valley) Tintara (McLaren Vale) Torbreck Vintners (Barossa) Trentham Estate (Murray Darling) Tulley Wells (Upper Goulburn) Turkey Flat Vineyards (Barossa) Twelve Staves Wine Company (McLaren Vale) Uleybury Wines (Adelaide Zone) Valhalla Wines (Rutherglen) Valley Wines (Swan District) Varrenti Wines (Grampians) Veritas (Barossa) Veronique (Barossa) Vinaceous (Various) Vinifera Wines (Mudgee) Vino Italia (Swan Valley) Vinrock (McLaren Vale) Vintara (Rutherglen) Virgara Wines (Adelaide Plains) Wallington Wines (Cowra) Western Range Wines (Perth Hills) White's Vineyard (Swan Valley) Willunga 100 Wines (McLaren Vale) Windy Creek Estate (Swan Valley) Winter Creek Wine (Barossa) Wirra Wirra (McLaren Vale) Witches Falls Winery (Granite Belt) Woodstock (McLaren Vale) Yaldara (Barossa) Yalumba Wine Company (Barossa) Yangarra Estate (McLaren Vale) Yelland and Papps (Barossa) Zitta Wines (Barossa) Zonte's Footstep (Langhorne Creek)

GRENACHE GRIS

This is the grey (really pink) version of Grenache. It is used in Southern France to make rose or white wines.

- Spinfex wines (McLaren Vale)

GRUNER VELTLINER

Synonyms: Gruner, Weissgipfler, Veltilini, Vetlinski Zelene, Zoldeveltelini

Maturity group: Early maturing, Wine Type: Moderate bodied white wine with high acidity.

References: C, H, O

This white variety is the basis of the Austrian wine industry, and is used elsewhere in Eastern Europe. While the variety was formerly used for producing everyday drinking wines in Austria modern winemakers are able to produce a range of attractive wines, sparkling, dry whites and sweet whites. There are unrelated varieties Fruhroter Veltliner and Roter Veltliner.

There is some interest in establishing the variety in Australia, and it would probably do well in most of the cooler Australian wine regions.

- Lark Hill Winery (Canberra District)

HARSLEVELU

Maturity group: This variety ripens just before midseason. Wine Type: Very full bodied wines with low acidity.

References: C, D, K, H, O

This white wine variety is used to make aromatic sweet white wines in the Tokaj Region of Hungary. It is second to Furmint with which it is blended to make Tokaji. In Australia it is being tried on both sides of the continent.

- Briery Estate, (Perth Hills) Eling Forest Estate (Southern Highlands) Noorinbee Selection Vineyards (Gippsland)

JACQUEZ

Synonyms: Black Spanish, Blue French, El Paso, Troya

References: K

This variety is believed to arisen as a natural hybrid of the European vine Vitis vinifera and an American species Vitis aestivalis. Like many such hybrids it has the virtues of vigour and resistance to pests and diseases, but the wines are often oddly flavored. Kerridge says "The variety may also be suitable for producing an unfermented grape juice resembling blackcurrant juice."

- Riversands (Queensland Zone)

KERNER

Maturity group: Early maturing Wine Type: Very light bodied wines with high acidity.

References: C, D, H, O

Kerner is a crossing of Trollinger and Riesling. It is popular in Germany and ripens early enough to be used in England. Kerner yields well but the quality of the wine is not as highly regarded as the wines produced from Riesling.

Curiously, this wine makes gold medal winning wines in Robinvale where the climate is decidedly un-English.

- Blown Away (McLaren Vale) Robinvale Wines (Murray Darling)

References: C, D, H, O

KYOHU

This is a Japanese variety, a cross with parents Campbell and Centennial. It is used as a table grape and also occasionally for wine. The berries are large and dark skinned.

Bella Ridge, the sole Australian producer uses Kyohu to make a semisweet white wine.

- Bella Ridge Estate (Swan District)

LAGREIN

Synonyms: Lagrain, Lagarino

Wine Type: Wines with super heavy body and with moderate acidity.

References: C, D, H, O

Lagrein is a promising wine variety from the Trentino-Alto Aldige region of North West Italy. The range of flavours ascribed to the variety includes bitter cherries and sour plums, or is it sour cherries and bitter plums. Don't worry there is also the flavour of chocolate in there as well. It can be used to make light reds as well as full bodied reds. It is also used for some impressive roses.

There is a small but growing group of Lagrein enthusiasts in Australia, who have been talking up the potential of the variety.

- Amietta Vineyard (Geelong) Bogie Man Wines (Strathbogie Ranges) Brown Brothers (King Valley) Chalmers (Murray Darling) Cobaw Ridge (Macedon Ranges) Domain Day (Barossa) Hartz Barn Wines (Eden Valley) Heartland Wines (Limestone Coast) Karanto Vineyards (Langhorne Creek) King River Estate)King Valley) Lazzar Wines (Mornington Peninsula) Piako Vineyards (Murray Darling) Point Leo Road Vineyard (Mornington Peninsula) Ridgemill Estate (Granite Belt) Rossiters (Murray Darling) Sugarloaf Ridge (Southern Tasmania) Zonte's Footstep (Langhorne Creek)

LAMBRUSCO MAESTRO

References: C D, H, O

Both a wine style and a grape variety, Lambrusco for most people means a light bodied, sweet red wine. These characteristics make Lambrusco a joke for many Australian winedrinkers who prefer big alcoholic wines. However Lambrusco is the name of a group of wine varieties, some of which make full bodied wines. One such is Trentham Estates' Lambrusco Maestro, which won a Gold medal a few years ago with a wine which would make many Aussie Shirazes look a bit woosy.

- Trentham Estate (Murray Darling)

LEMBERGER

Synonyms: Blaufrankisch, Limberger, Kekfrankos

Wine Type: Wines with moderate body and with high acidity.

References: C, D, O

No this is a cheese but one of the best German red wine varieties. It is also grown in Austria and Washington State. In Australia the only producer is Hahndorf Hill Winery which makes a straight varietal dry red and also blends it with Trollinger to make a very good dry rose.

- Hahndorf Hill (Adelaide Hills) Mac Forbes Wines (Yarra Valley)

MADELEINE ANGEVINE

This is an early ripening table and wine grape variety. The early ripening is useful in the UK, but curiously it is also grown in the Swan Valley where the climate is decidedly un-English. Harris Organic Wines (formerly LedaSwan) use it to make a fino style.

- Harris Organic Wines

MALBEC

Synonyms: Cot, Auxerrois, Pressac, Pied Rouge, Jacobain, Grifforin

Maturity group: This variety ripens just before midseason, Wine Type: Very full bodied wines with moderate acidity.

References: C, D, O

Malbec, under its French name of Cot, is one of the accepted varieties in the Bordeaux Appellation so it is debatable as to whether it qualifies as an alternative variety. Renewed interest in the variety, which was declining in importance until recently, justifies its inclusion.

Malbec is the mainstay of the famous 'black wines' of Cahors in South West France and also appears in several blended wines of the Loire.

Outside France Malbec's major stronghold is in Argentina where for many years it was the workhorse variety providing the material for everyday wines. Until the early 1990s the variety was subject of vine pull schemes, but modernisation of Argentina's wine industry has allowed Malbec's ability to make quality wines to be developed.

The variety is certainly less common in Australia than it was a couple of decades ago, but there is some evidence that it may be making a comeback. Malbec's role blending material with Cabernet has been supplanted by Merlot, and while there has always been a few varietal Malbecs around the variety was fading away.

At best varietal Malbecs can be rich and soft with a nice balance between fruit and tannins. They deserve more praise than they get.

- Adinfern (Margaret River) Alkoomi (Frankland River) Amicus (McLaren Vale) Andrew Peace Wines (Swan Hill) Ashton Hills (Adelaide Hills) Audrey Wilkinson (Hunter) Ballandeen Estate (Granite Belt) Barton Estate (Canberra) Beckett's Flat (Margaret River) Ben Potts Wines (Langhorne Creek) Bleasdale (Langhorne Creek) Bloodwood (Orange) Brave Goose Vineyard (Goulburn Valley) Bremerton (Langhorne Creek) Browns of Padthaway (Padthaway) Bungawarra (Granite Belt) Campbells Wines (Rutherglen) Cardinam Estate (Clare Valley) Carilley Estate (Swan Valley) Carpenteri

Vineyards (Swan Hill) Casa Freschi (Langhorne Creek) Celestial Bay (Margaret River) Chapel Hill (McLaren Vale) Cofield Wines (Rutherglen) Counterpoint Vineyard (Pyrenees) Cullen Wines (Margaret River) Delatite Winery (Upper Goulburn) Dinny Goonan Family Estate (Geelong) Disaster Bay Wines (South Coast Zone) Eldredge (Clare Valley) Emmetts Crossing Wines (Peel) Faber Vineyard (Swan Valley) Ferngrove Vineyards (Frankland River) Fryerstown Road Vineyard (Macedon Ranges) Gin Gin Wines (Queensland Coastal) Gipsie Jack (Langhorne Creek) Golden Ball (Beechworth) Governor's Choice Winery (Queensland Zone) Grassy Point Coatsworth Wines (Geelong) Gundowringla Wines (Alpine Valleys) Harcourt Valley (Bendigo) Higher Plane (Margaret River) Hills View (McLaren Vale) Joadja Vineyards (Southern Highlands) Karatta Wine (Robe) Kingsdale Wines (Southern New South Wales Zone) Kongwak Hills Winery (Gippsland) Lake Breeze (Langhorne Creek) McCuskers Vineyard (Perth Hills) McHenry Hohnen (Margaret River) Millbrook Winery (Perth Hills) Minnow Creek (McLaren Vale) Mount Charlie Winery (Macedon Ranges) Munari (Heathcote) Noorinbee Selection Vineyards (Gippsland) Nowra Hill Vineyard (Shoalhaven Coast) Olssens of Watervale (Clare Valley) Peter Lehmann (Barossa) Pondalowie (Bendigo) Pyren Vineyard (Pyrenees) Riverbank Estate (Swan Valley) Robertson of Clare (Clare Valley) Rodericks (South Burnett) Rogues Lane Vineyard (Heathcote) Rowanston on the Track (Macedon Ranges) Rudderless Wines (McLaren Vale) Settlers Ridge (Margaret River) Sevenhill Wines (Clare Valley) Skillogalee (Clare Valley) Stone Bridge Wines (Mount Lofty Ranges Zone) Summit Estate (Granite Belt) Sussanah Brook Wines (Swan District) Tahbilk (Nagambie Lakes) Tamborine Estate Wines (Queensland Coastal) Tamburlaine (Hunter) Tatachilla (McLaren Vale) Temple Bruer (Langhorne Creek) The Islander Estate Vineyards (Kangaroo Island) The Silos Estate (Shoalhaven Coast) Toowoomba Hills Estate (Queensland Zone) Victory Point Wines (Margaret River) Virgara Wines (Adelaide Plains) Warraroong Estate (Hunter) Waterwheel Wines (Bendigo) Wendouree (Clare Valley) Western Range Wines (Perth Hills) Wills Domain Vineyard (Margaret River) Windy Ridge Vineyard (Gippsland) Winya Wines (Queensland Zone) Wombat Lodge (Margaret River) Woodlands (Margaret River) Woodstock (McLaren Vale) Zonte's Footstep (Langhorne Creek)

MALIAN

References: This unique variety does not appear in standard references. The best source of information about Malian is the Cleggett website at www.cleggettwines.com.au

This variety arose as a sport of a Cabernet sauvignon vine at the Cleggett vineyard at Langhorne Creek. Cuttings from the vine were cultivated and the variety is now registered. The vine is identical to Cabernet sauvignon, but the berries are bronze-coloured. A further sport of Malian gave rise to the white wine variety Shalastin. Cleggett makes an early picked dry rose and a late picked rose style wine from a small planting of Malian.

- Cleggett (Langhorne Creek)

MALVASIA

Synonyms: Malmsey, Malvoise, Uva Greca, Cagazal, Rojal, Subirat, Monemvasia

Maturity group: This variety ripens just before midseason. Wine Type: Wines with moderate body and with low acidity.

References: C, D, O

Malvasia is not really a single variety but it is described in The Oxford Companion to wine as 'a complex web of varieties.' This situation no doubt arose because of Malvasia's origins in ancient Greece and its subsequent spread to most wine countries throughout the Mediterranean. The uses to which Malvasia is put are equally broad. On the island of Madiera it is used for the fortified Malmsy. In Italy white forms of Malvasia are blended with Trebbiano, it adds to the blend of Frascati, and red forms are blended with Sangiovese. Malvasia is aromatic, but not quite as strong as Muscat. In Australia Ermes Estate blend Malvasia with Riesling.

- Ermes Estate (Mornington Peninsula)

MAMMOLO

References: C, D, O

This is a highly perfumed red wine variety from Central Italy. Its name, which means 'violets' in Italian, is a reference to its characteristic perfume. In Italy Mammolo is used in small quantities in blends with other red varieties.

- Noorinbee Selection Vineyards (Gippsland)

MARSANNE

Synonyms: Avilleran, Grosse Roussette, Ermitage Blanc

Maturity group: This variety ripens in midseason, Wine Type: Full bodied wines with low acidity.

There has been renewed interest in Marsanne in Australia over the past decade. For many years the best known example was from Chateau Tahbilk in the Nagambie Lakes region of Central Victoria. This variety is a native to the Hermitage area in the Rhone Valley where it is regarded as a minor variety.

Wines made with this variety improve markedly with bottle age. It is less common for white wines to improve with bottle age, but this is an exception.

48 VARIETIES • Marsanne

With age wine develops a beautiful golden colour and the flavour rounds out to something resembling baked apples.

References: C, D, H, K, O

> Akrasi Wines (Central Victoria Zone) All Saints Estate (Rutherglen) Angullong Wines (Orange) Arlewood Estate (Margaret River) Banca Ridge (Granite Belt) Barwon Ridge Wines (Geelong) Battely Wines (Beechworth) Beechtree Wines (McLaren Vale) Beelgara Estate (Riverina) Belgrave Park Winery (South Coast Zone) Bents Road (Granite Belt) Bethanga Ridge) Bianchet (Yarra Valley) Botobolar (Mudgee) Brumfield (Yarra Valley) Bullers Calliope (Rutherglen) Calais Estate (Hunter) Cape Horn Vineyard (Goulburn Valley) Cape Mentelle (Margaret River) Charlies Estate Wines (Swan Valley) Churchview Estate (Margaret River) Connor Park (Bendigo) Coombe Farm Vineyard (Yarra Valley) Crane Wines (South Burnett) Cypress Post (Granite Belt) D'Arenberg (McLaren Vale) Dalfaras (Nagambie Lakes) Del Rios (Geelong) Djinta Djinta (Gippsland) Gilligan (McLaren Vale) Goulburn Terrace (Nagambie Lakes) Growlers Gully (Upper Goulburn) Hanging Rock Winery (Macedon Ranges) Happs (Margaret River) Harman's Ridge Estate (Margaret River) Heathcote Winery (Heathcote) Hidden Creek (Granite Belt) Honey Moon Vineyard (Adelaide Hills) Jarrets of Orange (Orange) Jinks Creek Winery (Gippsland) Jones Winery and Vineyard (Rutherglen) Kooroomba (Queensland Zone) Lillian (Pemberton) Lindenton Wines (Heathcote) Little's Winery (Hunter) Little River Wines (Swan Valley) M. Chapoutier Australia (Mount Benson) Mary Byrnes Wines (Granite Belt) McHenry Hohnen (Margaret River) McIvor Creek (Heathcote) McIvor Estate (Heathcote) McPherson Wines (Nagambie Lakes) Melaleuca Grove (Upper Goulburn) Mitchelton (Nagambie Lakes) Monument Vineyard (Central Ranges Zone) Mount Burrumboot Estate (Heathcote) Munari (Heathcote) Murray Street Vineyard (Barossa) Penny's Hill (McLaren Vale) Peppin Ridge (Upper Goulburn) Pfeiffer Wines (Rutherglen) Piggs Peake Winery (Hunter) Poachers Ridge Vineyards (Mount Barker) Ravensworth Wines (Canberra) Rimfire Vineyards (Darling Downs) Riverina Estate Wines (Riverina) Rutherglen Estates (Rutherglen) Saint Derycke's Wood (Southern Highlands) Seppelt Great Western (Grampians) Sirromet (Queensland Coastal) Spinifex (Barossa) Stanton Estate (Queensland Zone) Steinborner Family Vineyards (Barossa) Stone Ridge (Granite Belt) Stumpy Gully (Mornington Peninsula) Summit Estate (Granite Belt) Tahbilk (Nagambie Lakes) Tallarook Wines (Upper Goulburn) Tamburlaine (Hunter) Tawonga Vineyard (Alpine Valleys) Tempus Two (Hunter) Terra Felix (Upper Goulburn) Three Moon Creek (Queensland Zone) Torbreck Vintners (Barossa) Tulloch (Hunter) Turkey Flat Vineyards (Barossa) Valhalla Wines (Rutherglen) Wandoo Farm (Central Western Australian Zone) Wanted Man (Heathcote) Waratah Vineyard (Queensland Zone) Warrabilla Wines (Rutherglen) Winewood (Granite Belt) Wirruna Estate (North East Victoria) Witches Falls Winery (Granite Belt) Woop Woop Wines (McLaren Vale) Wrattonbully Vineyards (Wrattonbully) Yalumba Wine Company (Barossa) Yarra Glen (Yarra Valley) Yering Station (Yarra Valley) Yeringberg (Yarra Valley)

The codes for the References (C, D, H, K, O) are listed on page 9 of this book

MARZEMINO

Synonyms: Marzemino Gentile

Maturity group: Late ripening Wine Type: Full bodied wines with moderate acidity.

References: C, D, H, O

This is an Italian red wine variety of some potential. It is used mainly in Trentino, Veneto and Lombardy either for varietal wines or as blending material with Sangiovese and Barbera. Marzemino will probably play a minor role in the North East Victorian Italian varietals scene.

- Chrismont (King Valley), Michelini (Alpine Valleys)

MATARO

An Australian synonym for Mourvedre, which is sometimes still used in places of the official name.

MAVRODAPHNE

Wine Type: Wines with moderate body and with low acidity.

References: C, D, O

Mavrodaphne is an aromatic Greek variety that is most often used to make porty dessert style wines.

- Robinvale wines (Murray Darling)

MELON DE BOURGOGNE

Synonyms: Muscadet, Melon, Lyonnaise Blanche, Gamay Blanc a Feuilles Rondes

References: C, D, K, O

Maturity group: Early maturing, Group 3, Wine Type: Very light bodied wines with high acidity.

This variety is being pioneered in Australia by Garry Crittenden at his Mornington Peninsula winery. Melon, despite is full name, is best known for its role in the production of the wonderful Muscadet sur lie in the lower Loire Valley. It is no longer tolerated in Burgundy.

Muscadet sur lie is a dry wine produced exclusively from Melon. Winemaking involves long fermentation of the lees to extract plenty of flavour. The result is a crisp but flavourful dry white and the ideal accompaniment to seafood.

- Crittenden at Dromana (Mornington Peninsula) Whinstone Estate (Mornington Peninsula)

MERLOT

This variety is grown throughout Australia. As a mainstream variety it is beyond the scope of this book.

MEUNIER

Synonyms: Pinot Meunier, Dusty Miller, Miller's Burgundy, Gris Meunier, Farineux Noir, Mullertraube, Blanche Feuille, Schwartzriesling, Morillon Tacone

Maturity group: Very early maturity, Wine Type: Very full bodied wines with high acidity.

References: C, D, H, K, O

Meunier is one of three varieties used for Champagne, the others being Pinot Noir and Chardonnay. It is also used to make dry red wines of considerable character. Its most common synonym is Pinot meunier reflecting its close relationship with Pinot noir. Meunier means 'miller' and this is believed to have originated because the leaves of the vine have many fine hairs on their undersides and hence they seem to have been dusted with flour. The varieties early ripening ability lead to its widespread use throughout northern France and Germany, to produce both sparkling and still wines.

In Australia there have been increased plantings of Meunier as grower are looking to replicate the varietal composition of Champagne, but there is also renewed interest in its use for producing dry reds and rose. Over many decades Bests at Great Western have produced an excellent dry red from this variety; for a long time it was a beacon for those who wanted a wine with good flavour without the intense colour, jamminess and huge body of its contemporaries.

- Baillieu Vineyard (Mornington Peninsula) Barringwood Park (Northern Tasmania) Bests (Grampians) Bluestone Lane (Mornington Peninsula) Bochara (Henty) Borrodell on the Mount (Orange) Box Stallion (Mornington Peninsula) Brandy Creek Wines (Gippsland) Centennial Vineyards (Southern Highlands) Charles Melton (Barossa Valley) Collina del Re (King Valley) Harcourt Valley (Bendigo) Henderson Hardie (King Valley) John Gehrig Wines (King Valley) Kinloch Wines (Upper Goulburn) Manton's Creek Vineyard (Mornington Peninsula) Meadowbank Estate (Southern Tasmania)

Montalto Vineyards (Mornington Peninsula) MorganField (Macedon Ranges) Mount Macedon Winery (Macedon Ranges) Mt Billy (Southern Fleurieu) Rahona Valley Vineyard (Mornington Peninsula) Red Hill Estate (Mornington Peninsula) Starvedog Lane (Adelaide Hills) Symphonia (King Valley) Winbirra Vineyard (Mornington Peninsula) Woodonga Hill (Hilltops)

MONASTRELL

This is the Spanish name for Mourvedre

MONDEUSE

Synonyms: Monduese Noir, Gros Rouge de Pays, Grande Chetuan

Maturity group: Late ripening variety, Wine Type: Very full bodied wines with high acidity.

References: C, D, H, K, O

There has been some confusion about this variety. Some commentators suggested that it is the same variety as Refosco, and another theory was that it the same as Gros Syrah, the large berried form of Shiraz. The consensus now is that it is really a distinct variety. Mondeuse is grown in Savoy, where it is used as a varietal, but also in blends with Gamay and Pinot noir.

In Australia Mondeuse is used in blends with Shiraz and Cabernet (Brown Brothers) or just Shiraz (Chambers) for red wines capable of considerable bottle development. Plum and Cherry flavours and firm tannins are the virtues bought by Mondeuse to these wines.

- Brown Brothers (King Valley) Bullers Calliope (Rutherglen) Chambers Rosewood (Rutherglen) Eumundi Winery (Queensland Coastal) Hackersley (Geographe) St Petrox (Hunter) Symphony Hill Wines (Granite Belt)

MONTEPULCIANO

Synonyms: Cordisco, Morellone, Primaticcio, Uva Abbruzzi

Wine type: Full bodied low acid wines

References: C, D, O

First let's clear up some confusion about the name. Montepulciano is the name of both a grape variety and a town in Tuscany. This can cause problems as the wine and the town are not connected. There is a red wine called Vino Nobile di Montepulciano which is in fact made from the Sangiovese grape variety around the town of Montepulciano in Tuscany.

The grape variety Montepulciano is planted in Central and Southern Italy. It ripens late in the season and is thus unsuitable for the cooler northern regions of Italy. Montepulciano the grape variety has its most noteworthy expression is in the wine Montepulciano d'Abbruzzo from the mountainous region of Abruzzi on the Adriatic coast of Central Italy.

The wines made from Montepulciano are typically full bodied with deep colour, spicy and plummy flavours and high levels of tannin. They usually need some bottle age. You can pair them with hearty Italian cuisine or with sharp cheeses.

Australian wineries using Montepulciano have had encouraging results but it will be some time before we see the best wines from the variety

- Banrock Station (Riverland) Epsilon (Barossa) First Drop (Barossa) Oak Works (Riverland) Tscharke (Barossa)

MOSCATO

Moscato is the Italian word for Muscat. It can mean any of the several varieties of Muscat, especially Muscat a Petit Grains.

Over recent years low alcohol sweet or semi sweet wines under the name Moscato have become popular in Australia. These are typically low alcohol wines (6-10%) and are ideal as aperitifs. The pink or red equivalent is Zibibbo made from Muscat of Alexandria.

- Andrew Peace Wines (Swan Hill) Ballast Stone Estate (Currency Creek) Banrock Station (Riverland) Bellbrae Estate (Geelong) Box Stallion (Mornington Peninsula) Brown Brothers (King Valley) Bullers Beverford (Swan Hill) Casella (Riverina) Chalk Hill Winery (McLaren Vale) Chateau Tanunda (Barossa) Crittenden at Dromana (Mornington Peninsula) Cumulus Wines (Orange) De Bortoli (Riverina) Dominic Versace Wines (Adelaide Plains) Dos Rios (Swan Hill) Evans and Tate (Margaret River) Foxey's Hangout (Mornington Peninsula) Gapsted (Alpine Valleys) Glandore Estate (Hunter Valley) Gracebrook Vineyards (King Valley) Grant Burge (Barossa) Innocent Bystander (Yarra Valley) Jeanneret Wines (Clare Valley) Jindalee Estate (Geelong) Lake Breeze (Langhorne Creek) Lillypilly Estate (Riverina) Logan Wines (Mudgee) Pertaringa (McLaren Vale) Redbank Victoria (King Valley) Sam Miranda Wines (King Valley) Southern Highland Wines (Southern Highlands) T'Gallant (Mornington Peninsula) Tempus Two (Hunter) Terra Felix (Upper Goulburn) The Grapes of Ross (Barossa) The Wanderer (Yarra Valley) Tower Estate (Hunter)

Trentham Estate (Murray Darling) Westend Estate (Riverina) Wirra Wirra (McLaren Vale) Zilzie Wines (Murray Darling)

MOSCATO PARADISO

This variety was subject to some sleuthing by prominent Australian viticulturalist Dr Richard Smart. The variety was known in only one planting at Mudgee and was not known anywhere else. However it seems that the same variety was growing in the South Pacific Island of Funtina, under the name of "Saints wine." Its ultimate origin seems to be from a seedling in Chile. The only known planting in Australia now seems to be at Bago Vineyard in the Hastings River Region. They have produced a wine under the label of "Chanel Paradisa."

- Bago (Hastings River) Symphony Hill (Granite Belt)

MOURVEDRE

Synonyms: Mataro, Esparte, Balzac, Monastrell

Maturity group: Very late ripening variety, Wine Type: Wines with super heavy body and with high acidity.

References: C, D, H, K, O

Mourvedre is a Rhone variety which is becoming more popular in Australia, after a decline in the 1980s. In a way its fortunes have mirrored those of Grenache, a common blending partner. It was formerly more commonly known in Australia, and in California, as Mataro. Occasionally you will still find wines with that name on the label.

In France it is used a part of the blended red wines in the southern Rhone, often playing a supporting role to Grenache. Mourvedre is also an important variety in Provence, where it is used to make well structured roses, as well as serious reds, especially in the Appellation of Bandol.

Mourvedre is widely grown in Spain under the name of Monastrell. It is not so highly regarded in Spain, where its main role is to produce wines with high alcohol and tannin levels.

The most common use of this variety in Australia is in the increasingly popular Grenache, Shiraz Mourvedre (GSM) blends. These wines hark back to the 1960s when Australian dry red wine was labeled as claret and we weren't so fussy about what was on the label.

54 VARIETIES • Mourvedre

A couple of straight varietal Mourvedres made in Australia deserve special mention. Hewitson Old Garden Mourvedre is made from the fruit of 150-year-old vines in the Barossa Valley. Cascabel in the McLaren Vale produce a wine from this variety under the name of its Spanish synonym of Monastrell, reflecting winemaker Susana Fernandez's Spanish heritage.

- Andrew Peace Wines (Swan Hill) Annie's Lane (Clare Valley) Arakoon (McLaren Vale) Arimia Margaret River (Margaret River) B3 Wines (Barossa) Balthazar (Barossa) Barristers Block (Adelaide Hills) Belalie Bend (Southern Flinders Ranges) Bella Ridge Estate (Swan District) Blackbilly (McLaren Vale) Boireann (Granite Belt) Broken River Vineyards (Goulburn Valley) Bullers Beverford (Swan Hill) Burge Family Winemakers (Barossa) Canonbah Bridge (Western Plains) Cape Barren Wines (McLaren Vale) Cape Mentelle (Margaret River) Carpenteri Vineyards (Swan Hill) Cascabel (McLaren Vale) Chaperon Wines (Bendigo) Charles Melton (Barossa) Chateau Dorrien (Barossa) Clancy's of Conargo (Riverina) Clancy Fuller (Barossa) Claymore Wines (Clare Valley) Colonial Estate (Barossa) Connor Park (Bendigo) D'Arenberg (McLaren Vale) Deisen (Barossa) Di Fabio Estate (McLaren Vale) Diggers Bluff (Barossa) Diloreto Wines (Adelaide Plains) Disaster Bay Wines (South Coast Zone) Domain Barossa (Barossa) Eperosa (Barossa) Farosa Estate (Adelaide Plains) Geoff Merrill (McLaren Vale) Gibson Barossavale (Barossa) Gilligan (McLaren Vale) Gomersal Wines (Barossa) Gomersal Wines (Barossa) Grant Burge (Barossa) Greenstone Vineyard (Heathcote) Hamiltons Ewell Vineyards (Barossa) Hanging Rock Winery (Macedon Ranges) Happs (Margaret River) Henschke (Eden Valley) Hewitson (Barossa) Hutton Vale (Eden Valley) Izway Wines (Barossa) John Duval Wines (Barossa) Kabminye Wines (Barossa) Kaesler (Barossa) Kalleske Wines (Barossa) Kay Bros Amery (McLaren Vale) Kellermeister Wines (Barossa) Kennedy (Heathcote) Kilikanoon (Clare Valley) Kirrihill Estates (Clare Valley) Kirrihill (Adelaide Hills) Koltz (McLaren Vale) Kurtz Family Vineyards (Barossa) Landhaus Estate (Barossa) Langmeil (Barossa) Lanzthomson Wines (Barossa) Lillian (Pemberton) Limb Vineyards (Barossa) Lou Miranda Estate (Barossa) Magpie Estate (Barossa) Mansfield Wines (Mudgee) Marienberg (McLaren Vale) Marius Wines (McLaren Vale) Mary Byrnes Wines (Granite Belt) Massena Wines (Barossa) Maverick Wines (Barossa) McHenry Hohnen (Margaret River) McPherson Wines (Nagambie Lakes) Mitchell (Clare Valley) Mitchelton (Nagambie Lakes) Mount Appallan Vineyards (South Burnett) Mount Camel Ridge Estate (Heathcote) Mt Billy (Southern Fleurieu) Murray Street Vineyard (Barossa) Myattsfield Vineyard and Winery (Perth Hills) New Glory (Goulburn Valley) O'Donohue's Find (Riverland) Oak Dale Wines (Swan Hill) Olssens of Watervale (Clare Valley) Organic Vignerons Australia (Riverland) Paradine Estate (Queensland Zone) Pennyfield Wines (Riverland) Peter Lehmann (Barossa) Phoenix Estate (Clare Valley) Pikes (Clare Valley) Pyramid Gold (Bendigo) Pyramids Road Wines (Granite Belt) RBJ (Barossa) Redbox Perricoota (Perricoota) Reedy Creek (Northern Slopes Zone) Rockford (Barossa) Rusden Wines (Barossa) Rutherglen Estates (Rutherglen) Saltram (Barossa) Sarsfield Estate (Gippsland) Schild Estate Wines (Barossa) Schulz Vignerons (Barossa) Scorpiiion (Barossa) Seraph's Crossing (Clare Valley) Sieber Road Wines (Barossa) Silver Wings Winemaking (Goulburn Valley) Silverfox Wines (Perricoota) Sirromet (Queensland Coastal) Smallfry Wines (Barossa) Sons of Eden (Barossa) Soul Growers (Barossa) Spinifex (Barossa) Spook Hill Wines (Riverland) Springs Hill Vineyard (Fleurieu Zone) St Annes Vineyards

(Perricoota) St Hallett (Barossa) Sutherlands Creek Vineyard (Geelong) Tahbilk (Nagambie Lakes) Tallarook Wines (Upper Goulburn) Te-Aro (Barossa) Terra Felix (Upper Goulburn) Teusner (Barossa) The Old Faithful Estate (McLaren Vale) The Ritual (Peel) Tim Smith Wines (Barossa) Tin Shed Wines (Eden Valley) Torambre Wines (Riverland) Torbreck Vintners (Barossa) Turkey Flat Vineyards (Barossa) Valhalla Wines (Rutherglen) Veritas (Barossa) Veronique (Barossa) Vintara (Rutherglen) Wallington Wines (Cowra) Wendouree (Clare Valley) Wirra Wirra (McLaren Vale) Yaldara (Barossa) Yalumba Wine Company (Barossa) Yangarra Estate (McLaren Vale) Zitta Wines (Barossa) Zonte's Footstep (Langhorne Creek)

MULLER THURGAU

Synonyms: Riesling-Sylvaner, Rivaner

Maturity group: Very early maturity, Wine Type: Very light bodied wines with low acidity.

References: C, D, H, K, O

Muller Thurgau is a manmade variety, once commonly believed to be a Riesling Sylvaner cross, but DNA fingerprinting has shown its parents are actually Riesling and Chasselas de Courtillier. Whatever its parentage the variety gets a fairly bad press with many critics accusing it of producing dull wine. This is often done by way of comparison with Riesling, a fairly high hurdle for any variety to overcome. Nevertheless Muller Thurgau was widely planted in Germany and New Zealand, where its virtue of very early maturity made it quite popular.

Muller Thurgau's fortunes are now on the wane, and its future seems to be as a niche variety for regions where it is too cold to ripen other more attractive varieties. If there is hint of hot weather during ripening the acid levels in the grapes plummet and so does the prospect of making a good wine.

- Cross Rivulet (Southern Tasmania) Galafrey (Mount Barker) Green Valley Vineyard (Margaret River) Herons Rise (Southern Tasmania) Marions Vineyard (Northern Tasmania) Pieter van Gent (Mudgee) Snowy River Winery (Southern New South Wales Zone) Tarcoola Estate (Geelong) Whitehorse Wines (Ballarat) Wilmot Hills Vineyard (Northern Tasmania)

MUSCADELLE

Synonyms: Tokay, Sauvignon Vert

Wine Type: light bodied wines with low acidity, or sweet fortified wines.

References: C, D, H, K, O

This is the variety behind Australian Tokay wine. Traditional Tokay, from Hungary, uses the Furmit and Harsevelu grape varieties.

Muscadelle is used in Bordeaux as a minor blending partner along with Semillon and Sauvignon blanc, to make dry white wines, Sauternes and other sweet white wines. It is also used to make the famous semi-sweet Monbazillac white wines of the Bergerac region of SW France

Liqueur Tokay along with its close relation Liqueur Muscat, are two gems of Australian winemaking. The styles have been made for well over a century, principally in the Rutherglen and Glenrowan regions of NE Victoria. It was not until the 1970s that the grape variety previously known as "Tokay" was correctly identified as Muscadelle.

Both Muscat and Tokay liqueurs are made with similar methods. Semi-dried grapes are picked, partially fermented, fortified with grape spirit and then given various ageing treatments. Younger wines are back blended with older wines in a system similar to the solera system used in Spain for making sherry.

The Winemakers of Rutherglen classify their fortified wines into four groups depending largely on the age and therefore the richness and complexity of the wines. At the bottom is Rutherglen Tokay (or Muscat) then there is Classic Tokay, then Grand Tokay, then Rare Tokay. The prices rise accordingly, but you get more than your money's worth at every point.

The Muscadelle grape variety can be used alone or in blends to make dry white wines but its use for this purpose in Australia is fairly uncommon.

- All Saints Estate (Rutherglen) Anderson Winery (Rutherglen) Bailey's of Glenrowan (Glenrowan) Brown Brothers (King Valley) Bullers Calliope (Rutherglen) Campbells Wines (Rutherglen) Chambers Rosewood (Rutherglen) Charles Melton (Barossa) Cofield Wines (Rutherglen) Colonial Estate (Barossa) Crabtree Watervale Wines (Clare) Gehrig Estate (Rutherglen) Happs (Margaret River) Harris Organic Wines (Swan Valley) Hewitson (Barossa) Kladis Estate (Shoalhaven Coast) Morris (Rutherglen) Mount Prior (Rutherglen) Peter Lehmann (Barossa) Pfeiffer Wines (Rutherglen) Sevenhill Wines (Clare Valley) Smallfry Wines (Barossa) Stanton and Killeen Wines (Rutherglen) Talijancich (Swan Valley) Transylvania Winery (Southern New South Wales Zone) Whinstone Estate (Mornington Peninsula)

NEBBIOLO

Synonyms: Chiavennesca, Picutener, Spanna, Pugnet, Picoutener

Maturity group: Late ripening variety, Wine Type: Very full bodied wines with high acidity.

References: C, D, H, K, O

VARIETIES • Nebbiolo

Nebbiolo is a red grape variety whose home range is the Piedmonte region of Italy. It is the variety used to make the highly regarded Barolo and Barbaresco wines, named after the villages of their origin.

In Australia Nebbiolo is attracting some interest. Nebbiolo has the reputation of being a difficult grape to grow. Several growers have given up on the variety. It is a variety that needs to be matched closely with the appropriate microclimate. This task has yet to be completed. There seems to be a couple of producers trying Nebbiolo in many of the major Australian wine regions, but nowhere, with the possible exception of the Adelaide Hills has it been so successful as to induce neighboring growers to grow it.

Nebbiolo makes wine with a distinct brown colour. They are usually very long lived and often reward cellaring for a decade or so. The wines have rich flavours with a complex nose most often described as 'tar and roses.

- Aldinga Bay (McLaren Vale) Amulet Vineyard (Beechworth) Arrivo (Adelaide Hills) Aventine Wines (Granite Belt) Bacchus Hill (Sunbury) Ballandeen Estate (Granite Belt) Boireann (Granite Belt) Bowe Lees (Adelaide Hills) Brokenwood Wines (Hunter) Brown Brothers (King Valley) Capel Vale (Geographe) Carlaminda Estate (Geographe) Carlei Estate (Yarra Valley) Casa Freschi (Langhorne Creek) Ceres Bridge Estate (Geelong) Chain of Ponds (Adelaide Hills) Chestnut Hill Vineyard (Port Phillip Zone) Coriole (McLaren Vale) Counterpoint Vineyard (Pyrenees) Cow Hill (Beechworth) Delatite Winery (Upper Goulburn) Di Lusso Estate (Mudgee) Dromana Estate (Mornington Peninsula) Ferguson Falls Winery (Geographe) Fermoy Estate (Margaret River) First Drop (Barossa) Flat View Vineyard (Clare Valley) Flying Fish Cove (Margaret River) Galli Estate (Sunbury) Geographe Wines (Geographe) Glen Creek Wines (Upper Goulburn) Glenwillow Vineyard (Bendigo) Grove Estate Wines (Hilltops) Happs (Margaret River) Hidden Creek (Granite Belt) Indigo Wine Company (Beechworth) Jasper Hill (Heathcote) Kenton Hill (Adelaide Hills) La Cantina King Valley (King Valley) Lashmar (Kangaroo Island) Longview Vineyard (Adelaide Hills) Louee Wines (Mudgee) Luke Lambert Wines (Heathcote) Maglieri (McLaren Vale) McIvor Estate (Heathcote) Montefalco Vineyard (Porongurup) Monument Vineyard (Central Ranges Zone) Moondarra (Gippsland) Moppity Vineyards (Hilltops) Mount Cole Wineworks (Grampians) Mount Franklin Estate (Macedon Ranges) Mount Surmon (Clare Valley) Mount Towrong (Macedon Ranges) Oak Works (Riverland) Parish Hill Wines (Adelaide Hills) Pasut Family Wines (Murray Darling) Peel Ridge (Peel) Philip Lobley Wines (Upper Goulburn) Pizzini Wines (King Valley) Pokolbin Estate (Hunter) Primo Estate (Adelaide Plains) Protero (Adelaide Hills) Rigel Wines (Mornington Peninsula) Rojo Wines (Port Phillip Zone) Roselea Estate (Shoalhaven Coast) Rutherglen Estates (Rutherglen) Sandhurst Ridge (Bendigo) SC Pannell (McLaren Vale) Scaffidi Estate (Adelaide Hills) Serafino Wines (McLaren Vale) Sirromet (Queensland Coastal) Southern Highland Wines (Southern Highlands) Starvedog Lane (Adelaide Hills) Stefano Lubiano (Southern Tasmania) Stuart Wines (Heathcote) Swings & Roundabouts (Margaret River) Talunga (Adelaide Hills) Tannery Lane (Bendigo) Tar and Roses (Nagambie Lakes) Tenafeate Creek Wines (Adelaide Plains) Thorn-Clarke Wines (Barossa) Toppers Mountain (New England) Trandari (Hilltops) Trentham Estate (Murray Darling) Vinea Marson (Heathcote) Vineyard 28

(Geographe) Warrenmang Vineyard (Pyrenees) Witchmount Estate (Sunbury) Yalumba Wine Company (Barossa) Yandoit Hill Winery (Bendigo) Yering Station (Yarra Valley)

NEGROAMARO

Negro Amaro, Nigramaro

Wine type: very heavy bodied wine with high acidity.

References: C, D, H, O

This red wine variety is popular in the South of Italy, especially in Puglia (that's the heel of Italy's long leg.)

It is now attracting attention along with a host of other varieties from that region in the belief that the climate of many warmer Australian regions is closer to that of the South of Italy.

The name of the variety (a combination of the Italian words for *black* and *bitter*) give a clue to the flavour of the wine. It will take some marketing to get Australians to drink this wine. It doesn't lend itself to consumption without food.

- Chalmers (Murray Darling) Parish Hill Wines (Adelaide Hills)

NERO D'AVOLA

Synonyms: Calabrese

References: C, D, O

This is a Sicilian red wine variety that is regarded as being suitable for warmer Australian regions. The wines are deeply coloured but soft, perhaps less of a challenge than Negroamaro for Australian palates.

- Brown Brothers (King Valley) Chalmers (Murray Darling) Pertaringa (McLaren Vale)

ONDENC

Synonyms: Sercial, Irvine's White, Ondin, Ondent, Chalosse, Primai, Bequin

References: C, H, K, O

Ondenc is a rare white wine variety from South West France, where it is all but extinct. In Australia it was formerly used to make dry whites and base wine for sparkling wine. Langmeil winery in the Barossa Valley makes a sparkling

Ondenc Cuvee. Although this variety was grown under the name of Sercial, it is in fact different to the Sercial that is used on Madeira.

- Langmeil (Barossa) Seppelt (Grampians)

ORANGE MUSCAT

Synonyms: Moscato Fior d'Arrancio, Muscat primavis, Muscat de Jesus

References: C, H, O

This may or may not be a member of the Muscat family of grape varieties. In any case it is a minor variety that is grown in California and Australia. Orange Muscat is used to make sweet or dry wines, but it is difficult to sell against the wall of prejudice that says "muscat = sweet and sweet = unsophisticated."

As the name suggests this variety makes wines which have a strong, but not unpleasant, orange bouquet.

- All Saints (Rutherglen) Amulet (Beechworth) Brown Brothers (King Valley), Goorambath (Goulburn Valley) Monichino (Goulburn Valley) St Leonards (Rutherglen)

PALOMINO

Synonyms: Palomino fino, Palomino de Chipiona, Alban, Listan, Sweetwater, Paulo

Maturity group: Late ripening variety, Wine Type: Very light bodied wines with low acidity, and fortified sherry style wines.

References: C, D, H, K, O

Palomino is the dominant variety in the Spanish region of Jerez where it is used to make sherry. It is grown in Australia for a similar purpose, and is also used to make undistinguished white table wines. International agreements mean that the use of the term "sherry" for Australian wines is being phased out, but some of these wines are of incredible value.

- 919 Wines (Riverland) Angoves Winery (Riverland) Chambers Rosewood (Rutherglen) Kaesler (Barossa) Pennyweight Winery (Beechworth) Seppelt Great Western (Grampians) Settlement Wines (McLaren Vale)
-
-

- **The codes for the References (C, D, H, K, O) are listed on page 9 of this book**

PEDRO XIMENEZ

Synonyms: Pedro Jimenez, Pero Ximenen, Jerez

Maturity group: Early maturing, Group 3, Wine Type: Very light bodied wines with low acidity.

References: C, D, H, K, O

The other Sherry variety from Jerez, it is not nearly as common as Palomino. It is used to add sweetness to blends with Palomino and also on its own to make dark sweet viscous wines, both in Jerez and other regions in Southern Spain. Like Palomino it can also be used to make dry whites without much character.

In Australia it is used to make both fortified and table wines.

- Campbells Wines (Rutherglen) Charles Melton (Barossa) Chittering Valley Winery (Perth Hills) Harris Organic Wines (Swan Valley) Jones Winery and Vineyard (Rutherglen) Kellermeister Wines (Barossa) Lamonts (Central Western Australian Zone) Macaw Creek Wine (Mount Lofty Ranges Zone) Mansfield Wines (Mudgee) Phoenix Estate (Clare Valley) Settlement Wines (McLaren Vale) Sevenhill Wines (Clare Valley) Turkey Flat Vineyards (Barossa) Valley Wines (Swan District)

PETIT MANSENG

Synonyms: Manseng blanc Ichirota Zura Tipia

Wine Type: Full bodied white wines with moderate acidity.

References: C, D, H, O

Petit manseng is a high quality white wine variety which is native to South Western France in the regions of Jurancon and Pacherenc. In France it is used to make dry white wine with floral spicy fruit flavours and high acidity. It is also used to make sweet or semi sweet wines by allowing the grapes to shrivel on the vine in a process known as *passerillage*.

There is a Gros Manseng, but not in Australia.

While it is early days as yet there could be a bright future for this variety in Australia. It is more likely to be used for dry whites than sweetwer styles.

- 919 Wines (Riverland) Bochara (Henty) Chrismont (King Valley) Fighting Gully Road (Beechworth) Gapsted (Alpine Valleys) Mansfield Wines (Mudgee) Symphonia (King Valley)

PETIT MESLIER

References: H, O

Petit Meslier is a very rare white grape variety that is used in Champagne where it is valued for its ability to hold acid levels in warm years. The major champagne varieties, Chardonnay, Pinot noir and Meunier do not retain their acidity when the weather is too warm at ripening. The base wine used to make sparkling wine needs to have a high acidity.

Irvine Wines in the Eden Valley are the only Australian growers and producers using this variety. They planted a small amount in the mid 1980s which they originally used in a blend with Chardonnay based sparkling wine, but in the 1990s they decided to make a straight Petit Meslier. The result was a crisp sparkling wine with a bouquet of green apples. They have continued making a small commercial amount under the name Meslier Brut that sells out every year.

- Irvine (Eden Valley)

PETIT VERDOT

Synonyms: Verdot Rouge, Carmelin

Maturity group: Very late ripening variety, Wine Type: Wines with super heavy body and with high acidity.

References: C, D, H, K, O

Petit verdot is one of the lesser known and grown of the Bordeaux red grape varieties. As a late maturing variety it is losing popularity in the relatively cool Bordeaux region. The other four permitted varieties in the Bordeaux appellation are Cabernet sauvignon, Cabernet franc, Malbec and Merlot. These better known varieties are now widely planted in other regions of Europe and the New World. Petit verdot is less and less used in its native Bordeaux and has, until recently, been largely ignored elsewhere.

So why is Petit Verdot the poor cousin? In a way the answer lies in the vineyard, not the winery. This variety ripens later than the others and in Bordeaux, its traditional home, it may not ripen at all or produce such poor quality fruit as not to be worth the effort. In good (warm) years though, late ripening is the very attribute proves that nurturing the ugly duckling can pay off. It produces well coloured concentrated wines with a range of flavours, with violets being a common adjective used to describe its unique character. In Bordeaux it is used as a minor component (usually less than about five percent) of blends.

62 VARIETIES • Petit verdot

In Australia it is being planted in warmer areas where ripening is not such a problem.

The Australian tradition producing one hundred percent varietals is putting the acid test on Petit Verdot. It is probably fair to say that the results thus far have been very encouraging, but growers and makers are still at the pioneering stage. The characters too look for in varietal Petit verdots are intense colours, the wonderful fragrant nose, rich flavours and soft tannins.

- 1847 (Barossa) Aldinga Bay (McLaren Vale) Alkoomi (Frankland River) Anderson Winery (Rutherglen) Angoves Winery (Riverland) Arimia Margaret River (Margaret River) Arrowfield (Hunter) Ashton Hills (Adelaide Hills) Baarrooka (Strathbogie Ranges) Bago Vineyards (Hastings River) Ballast Stone Estate (Currency Creek) Banrock Station (Riverland) Barton Estate (Canberra) Beach Road (Langhorne Creek) Beechtree Wines (McLaren Vale) Beelgara Estate (Riverina) Bellarmine Wines (Pemberton) Bendigo Wine Estate (Bendigo) Blue Metal Vineyard (Southern Highlands) Boireann (Granite Belt) Bonneyview (Riverland) Bremerton (Langhorne Creek) Broomstick Estate (Margaret River) Brown Brothers (King Valley) Burnbrae Winery (Mudgee) Capercaillie (Hunter) Casella (Riverina) Celestial Bay (Margaret River) Ceravolo Premium Wines (Adelaide Plains) Chateau Mildura (Murray Darling) Clovely Estate (South Burnett) Cosham (Perth Hills) Counterpoint Vineyard (Pyrenees) Craneford (Barossa) Cullen Wines (Margaret River) D'Arenberg (McLaren Vale) De Beaurepaire Wines (Mudgee) De Bortoli (Riverina) Deakin Estate (Murray Darling) Di Fabio Estate (McLaren Vale) Disaster Bay Wines (South Coast Zone) Doctors Nose Wines (New England) DogRidge (McLaren Vale) Dowie Doole (McLaren Vale) Drakesbrook Wines (Peel) Drinkmoor Wines (Rutherglen) Duke's Vineyard (Porongurup) Eagle Vale (Margaret River) Eleven Paddocks (Pyrenees) Eumundi Winery (Queensland Coastal) Faber Vineyard (Swan Valley) Frog Rock (Mudgee) Fyffe Field (North East Victoria) Geddes Wines (McLaren Vale) Gemtree Vineyards (McLaren Vale) Geoff Hardy (McLaren Vale) Giant Steps (Yarra Valley) Gin Gin Wines (Queensland Coastal) Gipsie Jack (Langhorne Creek) Granite Ridge Wines (Granite Belt) Grey Sands (Northern Tasmania) Grove Estate Wines (Hilltops) Hackersley (Geographe) Hanging Rock Winery (Macedon Ranges) Harkaway Estate (Murray Darling) Haywards of Locksley (Strathbogie Ranges) Hugh Hamilton (McLaren Vale) Illalangi Wines (Riverland) Indigo Wine Company (Beechworth) Jamieson Estate (Mudgee) Jimbour Wines (Queensland Zone) Kahlon Estate Wines (Riverland) Kalleske Wines (Barossa) Keith Tulloch Wine (Hunter) Kevin Sobels Wines (Hunter) Kimber Wines (McLaren Vale) Kingston Estate (Riverland) Kirkham Estate (Sydney Basin) Kitty Crawford Estate (New England) Koppamura Wines (Wrattonbully) Kulkunbulla (Hunter) Kurtz Family Vineyards (Barossa) Lake Breeze (Langhorne Creek) Lankeys Creek Wines (Tumbarumba) Lazy River Estate (Western Plains) Leconfield (Coonawarra) Liebichwein (Barossa) Lilliput Wines (Rutherglen) Lillypilly Estate (Riverina) Little Wine Company (Hunter) Louee Wines (Mudgee) Mardia Wines (Barossa) Mason Wines (Granite Belt) McHenry Hohnen (Margaret River) Miles from Nowhere (Margaret River) Millbrook Winery (Perth Hills) Millers Dixons Creek Estate (Yarra Valley) Moama Wines (Perricoota) Mount Appallan Vineyards (South Burnett) Mount Burrumboot Estate (Heathcote) Mount Camel Ridge Estate (Heathcote) Mr Riggs Wine Company (McLaren Vale) Mudgee Growers (Mudgee) Mudgee Wines (Mudgee) Mulcra Estate Wines (Murray Darling) Nelwood Wines (Riverland) New

Glory (Goulburn Valley) Noorinbee Selection Vineyards (Gippsland) Nursery Ridge (Murray Darling) Olssens of Watervale (Clare Valley) Optimiste (Mudgee) Patrice Winery (North East Victoria) Penmara (Hunter) Pennyfield Wines (Riverland) Pertaringa (McLaren Vale) Petersons Glenesk Estate (Mudgee) Pettavel (Geelong) Phoenix Estate (Clare Valley) Piako Vineyards (Murray Darling) Pirramimma (McLaren Vale) Pyren Vineyard (Pyrenees) Ravens Croft Wines (Granite Belt) Red Cliffs (Murray Darling) Red Mud (Riverina) Remarkable View Winery (Southern Flinders Region) Riverbank Estate (Swan Valley) Roberts Estate (Murray Darling) Robertson of Clare (Clare Valley) Robyn Drayton (Hunter) Romantic Vineyard (Pyrenees) Rookery Wines (Kangaroo Island) Roslyn Estate (Southern Tasmania) Salena Estate (Riverland) Salisbury Winery (Murray Darling) Salomon Estate (Currency Creek) Sam Miranda Wines (King Valley) Sanguine Estate (Heathcote) Seven Mile Vineyard (Shoalhaven Coast) Seven Ochres Vineyard (Margaret River) Sevenoaks Wines (Hunter) Sigismondi Estate Wines (Riverland) Sirromet (Queensland Coastal) Smallfry Wines (Barossa) Smithbrook (Pemberton) St Mary's (Penola) St Michael's Vineyard (Heathcote) Staunton Vale Vineyard (Geelong) Stevens Brook Estate (Perricoota) Summit Estate (Granite Belt) Tall Poppy (Murray Darling) Talunga (Adelaide Hills) Tamburlaine (Hunter) Temple Bruer (Langhorne Creek) Tenafeate Creek Wines (Adelaide Plains) The Pawn Wine Company (Langhorne Creek) Thorn-Clarke Wines (Barossa) Three Moon Creek (Queensland Zone) Tizzana Winery (South Coast Zone) Trahna Rutherglen Wines (Rutherglen) Trentham Estate (Murray Darling) Tulloch (Hunter) Uleybury Wines (Adelaide Zone) Victory Point Wines (Margaret River) Vintara (Rutherglen) Wallington Wines (Cowra) Waratah Vineyard (Queensland Zone) Waurn Ponds Estate (Geelong) Wenzel Family Wines (Langhorne Creek) Westlake Vineyards (Barossa) Whale Coast Wines (Southern Fleurieu) Wildwood (Sunbury) Williams Springs Road (Kangaroo Island) Wills Domain Vineyard (Margaret River) Windowrie Estate (Cowra) Wirra Wirra (McLaren Vale) Wombat Lodge (Margaret River) Woodstock (McLaren Vale) Wordsworth Wines (Geographe) Yaldara (Barossa) Yalumba Wine Company (Barossa) Yarran (Riverina) Zilzie Wines (Murray Darling) Zonte's Footstep (Langhorne Creek)

PICOLIT

Wine type: A white wine variety used mainly for sweet wines

References: C, H, O

This is a rare northeastern Italian white wine variety. It is used for dry whites and also late picked and *passito* styles. The Italian wines produced from this variety achieve great heights and astronomical prices. Like other Italian varieties Picolit is under active scrutiny here and we could see some interesting wines in the next few years. Di Lusso Estate has released several vintages of a dessert wine made from Picolit

- Di Lusso Estate (Mudgee) McAdams Lane (Geelong) Moondarra (Gippsland) Pizzini Wines (King Valley) Turners Crossing Vineyard (Bendigo) Witchmount Estate (Sunbury)

PICPOUL

Synonyms: Picapoll, Piquepoul blanc, Languedocien, Avilla

Wine Type: White wines with moderate body and with high acidity.

References: C, D, O

Picpoul is a white wine variety from the Languedoc Region in France. It is used to make highly acid wines. *Pique poul* means lip stinger in French. There is a dark skinned variety Picpoul noir. Picpouls is used to make the base wine for vermouth in the Languedoc region.

- Symphony Hill Winery (Granite Belt)

PINOT BLANC

Synonyms: Pinot bianco, Clevner, Chasselas Dorato, Borgogna bianco, Pineau blanc, Beli pinot, Weisser Rulander, Weisser Burgunder

Wine Type: Wines with moderate body and with high acidity.

References: C, D, H, O

This variety is most often associated with the Alsace region in France as well as the Italian regions of Trentino-Alto Adige, Veneto, Friuli and Lombardy. The variety is also quite widely grown in Germany, Austria and Eastern Europe.

Pinot Blanc bears some resemblance to Chardonnay in the vineyard but the wines tend to have higher acid. This higher acidity is valued for making base wines for sparkling wines, and when Pinot blanc is blended with other varieties.

Varietal Pinot blancs are uncommon but they can be attractive crisp wines.

- Amulet Vineyard (Beechworth) Hoddles Creek Estate (Yarra Valley) jb Wines (Barossa) Lindenderry (Mornington Peninsula) Mosquito Hill Wines (Southern Fleurieu) St Matthias (Northern Tasmania)

PINOT GRIS

Synonyms: Pinot grigio, Pinot Burot, Tokay d'Alsace, Tokayer, Malvoise, Gris Cordelier, Klevanjka, Rulander, Fromentot, Grauer Burgunder, Grauer Monch, Szurkebarat

Maturity group: Very early maturity, Wine Type: Full bodied wines with moderate acidity.

References: C, D, H, K, O

Over the past decade one of the hottest new white wine varieties in Australia is Pinot gris. There are now well over three hundred Australian wineries using this variety. Pinot grigio has done this despite a couple of marketing hurdles which would have stifled any ordinary variety.

A major problem is that there are two names for the variety in common use. The names Pinot grigio and Pinot gris are the both correctly used for the same variety. This is quite confusing in the marketplace. The words grigio and gris mean 'grey' in Italian and French respectively. The word 'pinot' refers to the pine cone shaped bunches of grapes characteristic to this group of varieties, which also includes Pinot noir, Pinot blanc and Pinot meunier.

In Australia, winemakers, marketers and wine writers haven't really sorted out which name to call the variety. Some have dodged the issue and refer to "Pinot G." The logical thing to do would be to call lighter styles grigio, as they are more akin to Italian wines, while using gris for the fuller styles more like the French.

This variety closely is related to the much better known Pinot Noir variety and is believed to be a mutation of Pinot Noir. In fact in the vineyard Pinot grigio is difficult to distinguish from its putative ancestor until the berries ripen when those of the Grigio will have much less pigment. There is another related variety, Pinot Blanc which has little or no pigment in the berries.

There is considerable clonal variation within the variety. This means that vines and wines from different vineyards can be quite different. Wine writer Jancis Robinson says that the variety hardly knows if it is a dark or a light grape

In Europe the variety is widely planted. Given the inherent variation and geographic dispersion it is hardly surprising that a wide range of wine styles are produced from it. In Alsace under the name of Tokay d'Alsace, it produces a rich, floral, dry, almost oily wine. In Northern Italy the Pinot Grigio's are light and sometimes even spritzig. Under the name of Rulander in Germany it produces dry wines somewhat similar to white burgundies.

Pinot gris has attracted serious interest in Australia only over the past couple of decades. It is now producing some remarkable wines in regions such as the Mornington Peninsula, Geelong and the Adelaide Hills.

There is a great deal of experimentation with the variety and it may take several more years before the optimal combination of terroir and winemaking technique emerges. In the meantime some great wines are already available for those who are looking for new experience.

The style varies from light bodied and fairly straight forward to rich and complex wines that are almost overwhelming in their voluptuousness. This is

66 VARIETIES • Pinot gris

one occasion when reading the back label or some tasting notes about a particular wine before buying is worthwhile.

- Ada River (Gippsland) Adina Vineyard (Hunter Valley) Alta Wines (Adelaide Hills) Amadio (Adelaide Hills) Amulet Vineyard (Beechworth) Anderson Winery (Rutherglen) Angoves Winery (Riverland) Angullong Wines (Orange) Annapurna (Alpine Valleys) Artwine (Clare Valley) Ashton Hills (Adelaide Hills) Austin's Wines (Geelong) Baie Wines (Geelong) Baillieu Vineyard (Mornington Peninsula) Ballast Stone Estate (Currency Creek) Banks Road (Geelong) Barmah Park Wines (Mornington Peninsula) Barringwood Park (Northern Tasmania) Barrymore Estate (Mornington Peninsula) Barton Estate (Canberra) Barwang (Hilltops) Basalt Ridge (Macedon Ranges) Bass Fine Wines (Northern Tasmania) Bass River (Gippsland) Battunga Vineyards (Adelaide Hills) Bay of Fires (Northern Tasmania) Bayview Estate Winery (Mornington Peninsula) Beelgara Estate (Riverina) Bellarine Estate (Geelong) Bergamin Wines (King Valley) Berton Vineyards (Riverina) Bimbadgen Estate (Hunter Valley) BK Wines (Adelaide Hills) Blackford Stable Wines (Adelaide Hills) Blue Metal Vineyard (Southern Highlands) Boat O'Craigo (Yarra Valley) Bockman (Adelaide Hills) Boggy Creek Vineyards (King Valley) Boyntons Feathertop (Alpine Valleys) Brammar Estate (Yarra Valley) Brandy Creek Wines (Gippsland) Brokenwood Wines (Hunter Valley) Brown's Farm Winery (Hunter Valley) Brown Magpie Wines (Geelong) Bulong Estate (Yarra Valley) Cape Banks (Limestone Coast) Cape Bernier Vineyard (Southern Tasmania) Casella (Riverina) Caught Redhanded (Adelaide Zone) Centennial Vineyards (Southern Highlands) Ceravolo Premium Wines (Adelaide Plains) Ceres Bridge Estate (Geelong) Chain of Ponds (Adelaide Hills) Chapel Hill (McLaren Vale) Chartley Estate (Northern Tasmania) Chateau Mildura (Murray Darling) Chateau Tanunda (Barossa Valley) Chrismont (King Valley) Clearview Estate Mudgee (Mudgee) Cleveland (Macedon Ranges) Cloudbreak Wines (Adelaide Hills) Clyde Park Vineyard (Geelong) Coldstream Hills (Yarra Valley) Constable Vineyards (Hunter Valley) Coombe Farm Vineyard (Yarra Valley) Crittenden at Dromana (Mornington Peninsula) Cumulus Wines (Orange) Curlewis Winery (Geelong) Curly Flat (Macedon Ranges) Cuttaway Hill Estate (Southern Highlands) Dal Zotto Estate (King Valley) Dalfaras (Nagambie Lakes) Darling Park (Mornington Peninsula) David Hook Wines (Hunter Valley) De Lisio Wines (McLaren Vale) Delatite Winery (Upper Goulburn) Deviation Road (Adelaide Hills) Di Lusso Estate (Mudgee) Doc Adams (McLaren Vale) Dos Rios (Swan Hill) Dromana Estate (Mornington Peninsula) Eagles Rise (Geelong) Elgee Park (Mornington Peninsula) Ellender Estate (Macedon Ranges) Ermes Estate (Mornington Peninsula) Farago Hill (Southern Highlands) Five Sons Estate (Mornington Peninsula) Foxey's Hangout (Mornington Peninsula) Freeman Vineyards (Hilltops) French Island Vineyards (Mornington Peninsula) Frog Rock (Mudgee) Frogmore Creek Vineyard (Northern Tasmania) Galli Estate (Sunbury) Galvanized Wine Group (McLaren Vale) Gapsted (Alpine Valleys) Ghost Rock (Northern Tasmania) Giant Steps (Yarra Valley) Glen Creek Wines (Upper Goulburn) Glenguin (Hunter Valley) Goorambath (Glenrowan) Gracebrook Vineyards (King Valley) Granite Ridge Wines (Granite Belt) Grant Burge (Barossa Valley) Grey Sands (Northern Tasmania) GrumbleBone Estate Vineyard (Southern Highlands) Hahndorf Hill (Adelaide Hills) Hanging Rock Winery (Macedon Ranges) Harlow Park Estate (Mornington Peninsula) Haselgrove (McLaren Vale) Hazyblur Wines (Kangaroo Island) Heartland Wines (Limestone Coast) Henderson Hardie (King Valley) Henley Hill (Yarra Valley) Henschke (Eden Valley) Herbert Vineyard (Mount Gambier) Heytesbury Ridge

VARIETIES • Pinot gris

(Geelong) Hillstowe (Adelaide Hills) Hillwood Vineyard (Northern Tasmania) Hoddles Creek Estate (Yarra Valley) HPR Wines (Mornington Peninsula) Hungerford Hill (Hunter Valley) Ibis Wines (Orange) Indigo Wine Company (Beechworth) Inneslake (Hastings River) Innocent Bystander (Yarra Valley) Iron Pot Bay Wines (Northern Tasmania) Irvine (Eden Valley) Jinglers Creek (Northern Tasmania) Jinks Creek Winery (Gippsland) Jones Road (Mornington Peninsula) Karanto Vineyards (Langhorne Creek) Kellermeister Wines (Barossa Valley) Kilgour Estate (Geelong) Killara Estate (Yarra Valley) Kirrihill (Adelaide Hills) Knappstein Wines (Clare Valley) Kouark (Gippsland) Kreglinger Estate (Mount Benson) Kurabana (Geelong) Lady Bay Winery (Southern Fleurieu) Lambert Vineyards (Canberra) Lark Hill Winery (Canberra) Laurellyn Wines (New England) Lazzar Wines (Mornington Peninsula) Leabrook Estate (Adelaide Hills) Lerida Estate (Canberra) Leura Park Estate (Geelong) Little Wine Company (Hunter Valley) Llangibby Estate (Adelaide Hills) Lochmoore (Gippsland) Logan Wines (Mudgee) Long Rail Gully (Canberra) Lou Miranda Estate (Barossa Valley) Louee Wines (Mudgee) Lowe Family Wines (Mudgee) Madew Wines (Canberra) Manton's Creek Vineyard (Mornington Peninsula) Marions Vineyard (Northern Tasmania) Maritime Estate (Mornington Peninsula) Mawson Ridge (Adelaide Hills) McAdams Lane (Geelong) McCrae Mist Wines (Mornington Peninsula) McLaren Wines (McLaren Vale) McVitty Grove (Southern Highlands) Meadowbank Estate (Southern Tasmania) Melross Estate (Pyrenees) Miceli (Mornington Peninsula) Michelini (Alpine Valleys) Millers Dixons Creek Estate (Yarra Valley) Milton Vineyard (Southern Tasmania) Minko (Southern Fleurieu) Monichino Wines (Goulburn Valley) Montalto Vineyards (Mornington Peninsula) Monument Vineyard (Central Ranges Zone) Moondarra (Gippsland) Moorilla Estate (Southern Tasmania) Moorooduc Estate (Mornington Peninsula) Morning Star Estate (Mornington Peninsula) Morning Sun Vineyard (Mornington Peninsula) Mount Ashby Estate (Southern Highlands) Mount Franklin Estate (Macedon Ranges) Mount Langi Ghiran Vineyards (Grampians) Mount Majura (Canberra) Mount Markey (Gippsland) Mount Pierrepoint Estate (Henty) Mount Surmon (Clare Valley) Mountadam (Eden Valley) Mt Samaria Vineyard (Goulburn Valley) Murdup Wines (Mount Benson) Mylkappa Wines (Adelaide Hills) Nalbra Estate (Geelong) Nazaaray (Mornington Peninsula) Nepenthe (Adelaide Hills) Neqtar Wines (Murray Darling) Nova Vita Wines (Adelaide Hills) Nugan Estate (King Valley) Nyora Vineyard and Winery (Gippsland) Oakridge (Yarra Valley) Oatley Wines (Mudgee) Ocean Eight Vineyard and Winery (Mornington Peninsula) Optimiste (Mudgee) Orchard Road (Orange) Outlook Hill (Yarra Valley) Paracombe Wines (Adelaide Hills) Paradigm Hill (Mornington Peninsula) Paramoor Wines (Macedon Ranges) Paringa Estate (Mornington Peninsula) Pasut Family Wines (Murray Darling) Paul Bettio (King Valley) Paxton (McLaren Vale) Peninsula Baie Wines (Geelong) Petaluma (Adelaide Hills) Pewsy Vale (Eden Valley) Phaedrus Estate (Mornington Peninsula) Piako Vineyards (Murray Darling) Pialligo Estate (Canberra) Pier 10 (Mornington Peninsula) Pike and Joyce (Adelaide Hills) Pikes (Clare Valley) Pipers Brook Vineyard (Northern Tasmania) Pirie Estate (Northern Tasmania) Piromit Wines (Riverina) Point Leo Road Vineyard (Mornington Peninsula) Politini (King Valley) Pooley Wines (Southern Tasmania) Pothana (Hunter Valley) Prancing Horse Estate (Mornington Peninsula) Primerano (King Valley) Primo Estate (Adelaide Plains) Prince Hill Wines (Mudgee) Printhie Wines (Orange) Provenance Wines (Geelong) Punt Road (Yarra Valley) Quealy (Mornington Peninsula) Quoin Hill (Pyrenees) Red Hill Estate (Mornington Peninsula) Redbank Victoria (King Valley) Redbox (Yarra Valley) (Yarra Valley) Ridgeview Wines (Hunter Valley) Rochford Wines (Yarra Valley) Rosevears Estate

(Northern Tasmania) Rowans Lane Wines (Henty) Ruane Winery (Southern Highlands) Sailors Falls Winery (Macedon Ranges) Sautjan Vineyards (Macedon Ranges) SC Pannell (McLaren Vale) Scorpo Wines (Mornington Peninsula) Scotchmans Hill (Geelong) Seaforth Vineyard (Mornington Peninsula) Seppelt Great Western (Grampians) Settlement Wines (McLaren Vale) Seville Estate (Yarra Valley) Shadowfax Vineyard and Winery (Geelong) Sidewood Estate (Adelaide Hills) Silverwaters Vineyard (Gippsland) Sirromet (Queensland Coastal) Snobs Creek Wines (Upper Goulburn) Somerbury Estate (Mornington Peninsula) South Channel Wines (Mornington Peninsula) Southern Highland Wines (Southern Highlands) Spring Vale Wines (Southern Tasmania) St Matthias (Northern Tasmania) Starvedog Lane (Adelaide Hills) Stefani Estate (Yarra Valley) Stefano Lubiano (Southern Tasmania) Stockman's Ridge (Central Ranges Zone) Stone Bridge Wines (Mount Lofty Ranges Zone) Stone Coast Wines (Wrattonbully) Stumpy Gully (Mornington Peninsula) Sugarloaf Ridge (Southern Tasmania) Sutherlands Creek Vineyard (Geelong) Symphonia (King Valley) Symphony Hill Wines (Granite Belt) T'Gallant (Mornington Peninsula) Tahbilk (Nagambie Lakes) Taltarni (Pyrenees) Tamar Ridge (Northern Tasmania) Tanjil Wines (Gippsland) Tar and Roses (Nagambie Lakes) Tarup Ridge Winery (Strathbogie Ranges) Taylors (Clare Valley) Te-Aro (Barossa Valley) Telgherry (Hunter Valley) Tempus Two (Hunter Valley) Ten Minutes by Tractor (Mornington Peninsula) The Cups Estate (Mornington Peninsula) The Garden Vineyard (Mornington Peninsula) The Lane (Adelaide Hills) The Pawn Wine Company (Langhorne Creek) Thorn-Clarke Wines (Barossa Valley) Three Willows Vineyard (Northern Tasmania) Tilbrook Estate (Adelaide Hills) Tim Adams (Clare Valley) Tomich Hill (Adelaide Hills) Toogoolah Wines (Orange) Toolangi Vineyard (Yarra Valley) Totino Wines (Adelaide Hills) Trentham Estate (Murray Darling) Tuck's Ridge (Mornington Peninsula) Tyrrells (Hunter Valley) Vale Creek Wines (Central Ranges Zone) Vale Vineyard (Mornington Peninsula) Velo Wines (Northern Tasmania) Vintina Estate (Mornington Peninsula) Walden Woods Farm (New England) Warburn Estate (Riverina) Waybourne (Geelong) Wellington (Southern Tasmania) Westend Estate (Riverina) Whinstone Estate (Mornington Peninsula) White Rock Vineyard (Northern Tasmania) Whyworry Wines (New England) Wild Cattle Creek Winery (Yarra Valley) Winbirra Vineyard (Mornington Peninsula) Winchelsea Estate (Geelong) Witchmount Estate (Sunbury) Wood Park (King Valley) Word of Mouth Wines (Orange) Wright Robinson of Glencoe (New England) Wyuna Park Vineyard (Geelong) Yabby Lake Winery (Mornington Peninsula) Yarra Burn (Yarra Valley) Yarra Ridge (Yarra Valley) Yaxley Estate (Southern Tasmania) Yellymong (Swan Hill) Yering Station (Yarra Valley) Zilzie Wines (Murray Darling) Zonte's Footstep (Langhorne Creek)

PINOT NOIR

This variety is grown in the cooler wine regions of Australia, especially in Tasmania and Victoria. As a mainstream variety it is outside the scope of this book.

PINOT MEUNIER

See Meunier

PINOTAGE

References: C, D, O

This red wine variety was bred in South Africa as a crossing of Pinot Noir and Cinsaut. It had the blessing, or curse, of being a high yielding variety and hence was widely planted and made gallons of not very good wine. But some enthusiasts decided to use some of the fruit from old bush vines and discovered that the variety is capable of making high quality and distinctive quality wine.

There has been Pinotage grown in New Zealand for a number of years and now a couple of wineries, in contrasting areas, have taken up the challenge in Australia

- Oak Works (Riverland) Whyworry Wines (New England)

PROSECCO

References: C, D, O

Prosecco is the name of a dry sparkling wine from Veneto in North East Italy as well as the variety that is used to produce it.

The variety is starting to catch on in Australia as a market for low alcohol, refreshing sparkling wines develops.

- Bogie Man Wines (Strathbogie Ranges) Box Grove Vineyard (Nagambie Lakes) Boyntons Feathertop (Alpine Valleys) Chrismont (King Valley) Dal Zotto Estate (King Valley) Mount Towrong (Macedon Ranges) Sam Miranda Wines (King Valley)

REFOSCO

Synonyms: Cagnina, Teran

Wine Type: Very full bodied wines with high acidity.

References: C, D, O

Refosco is a North Eastern Italian variety known for its acidic deeply coloured red wines. Like many Italian varieties there is some confusion about this variety and its close relatives, especially Mondeuse.

- Blanche Barkley (Bendigo) Daylesford Wine Company (Macedon Ranges)

REGENT

References: O

This red variety is a recent crossing with parentage of Sylvaner, Muller Thurgau and Chambourcin. Obviously the breeders were looking for disease resistance (Chambourcin has American ancestors) as well as a short time for maturity suitable for Germany. It is under trial in Queensland.

- Eumundi Winery (Queensland Coastal)

RIESLING

Riesling is a mainstream grape variety and outside the scope of this book.

RONDINELLA

Wine Type: Wines with moderate body and low acidity.

References: C, D, H, O

This red variety's main claim to fame is its role in the Valpolicella blend in the Veneto region of Italy where it is regarded as playing a support role to the more characterful Corvina. It is also used in the blend for the famous Amarone wine which is made from semi dried grapes. Freeman Vineyards are using Rondinella in a blend with Corvina for this purpose.

- Centennial Vineyards (Southern Highlands) Freeman vineyards (Hilltops)

ROTER VELTLINER

References: O

Roter Veltliner is an Australian variety used for wine as well as table grapes. Boat O'Craigo in the Yarra Valley uses it in small quantities to add body as well as almond and spice flavours to their Pinot gris.

- Boat O'Craigo (Yarra Valley)

ROUSSANNE

Synonyms: Bergeron, Barbarin, Picotin Blanc

Maturity group: This variety ripens in midseason, Wine Type: Full bodied wines with high acidity.

References: C, D, H, K, O

Roussanne is another Northern Rhone white wine variety that is undergoing a revival. In the Rhone it is used in blends with Marsanne for example in White Hermitage, it also one of thirteen varieties permitted in Chateauneuf de Pape. Roussanne is less popular than Marsanne because of its irregular yield, late ripening and susceptibility to powdery mildew. Nevertheless Roussanne has been adopted by the Rhone Rangers in California, perhaps because of its ability to produce wines with a unique aroma, often described as being like herb tea. It is also known to produce whites with great potential for aging.

In Australia the variety has only recently attracted much attention. However plantings are increasing and we are likely to see more Roussanne, either as a varietal or in blends with Marsanne and sometimes Viognier as well. Kabminye in the Barossa use this variety in a red blend in a similar way to Chateauneuf de Pape.

- All Saints Estate (Rutherglen) Arlewood Estate (Margaret River) Belgrave Park Winery (South Coast Zone) Belgravia Vineyards (Orange) Bellbrae Estate (Geelong) Box Grove Vineyard (Nagambie Lakes) Brown Brothers (King Valley) Campbells Wines (Rutherglen) Cape Mentelle (Margaret River) Cascabel (McLaren Vale) Catspaw Farm (Granite Belt) Chambers Rosewood (Rutherglen) Counterpoint Vineyard (Pyrenees) D'Arenberg (McLaren Vale) Djinta Djinta (Gippsland) Galvanized Wine Group (McLaren Vale) Giaconda (Beechworth) Gilligan (McLaren Vale) Goona Warra Vineyard (Sunbury) Growlers Gully (Upper Goulburn) Indigo Wine Company (Beechworth) Jerusalem Hollow (Margaret River) Kabminye Wines (Barossa) Lillian (Pemberton) Lowe Family Wines (Mudgee) Mary Byrnes Wines (Granite Belt) Massena Wines (Barossa) McHenry Hohnen (Margaret River) McIvor Estate (Heathcote) Mitchelton (Nagambie Lakes) Penny's Hill (McLaren Vale) Seppelt Great Western (Grampians) Snobs Creek Wines (Upper Goulburn) St Huberts (Yarra Valley) Tahbilk (Nagambie Lakes) Tallarook Wines (Upper Goulburn) Terra Felix (Upper Goulburn) Torbreck Vintners (Barossa) Turkey Flat Vineyards (Barossa) Waterwheel Wines (Bendigo) Wood Park (King Valley) Woop Woop Wines (McLaren Vale) Yangarra Estate (McLaren Vale) Yeringberg (Yarra Valley) Zonte's Footstep (Langhorne Creek)

RUBIENNE

Maturity group: Late ripening Wine Type: Full bodied with high acidity.

References: A new variety which has not yet been commercially released. A fact sheet is at www.csiro.au/proprietaryDocuments/PI_info_rubienne.pdf

This variety was bred in the 1970s by the CSIRO as a cross between the Spanish grape variety Sumoll and Cabernet Sauvignon. The intention was to produce a variety which would outperform Cabernet sauvignon in warmer irrigation districts. Vineyard trials, wine making and sensory evaluation are all encouraging. Commercial trials are being conducted by Orlando Wyndham.

RUBY CABERNET

References: C, D, H, K, O

This variety is a cross of Carignan with Cabernet Sauvignon, the intention being to breed a wine which could perform well in hot conditions.

Ruby Cabernet enjoyed a surge in popularity in Australia during the 1990s based on its capacity for high yields rather than for the quality of its wines. It is used for blended wines at the lower price ranges and for cask wines. In neither case is the blurb on the label likely to emphasize the contribution of Ruby Cabernet to the wine, if it is mentioned at all. It can also be used for fortified styles.

- Aldinga Bay (McLaren Vale) All Saints Estate (Rutherglen) Angoves Winery (Riverland) Blackwood Wines (Blackwood Valley) Campbells Wines (Rutherglen) Celestial Bay (Margaret River) Crane Wines (South Burnett) Crooked River Wines (Shoalhaven Coast) Djinta Djinta (Gippsland) Frogmore Creek Vineyard (Northern Tasmania) Granite Ridge Wines (Granite Belt) Grant Burge (Barossa) Monument Vineyard (Central Ranges Zone) Oceanview Estates (Queensland Coastal) Pier 10 (Mornington Peninsula) Richfield Estate (New England) Rimfire Vineyards (Darling Downs) Riversands Winery (Queensland Zone) Robinvale Wines (Murray Darling) Romavilla (Roma) Sevenhill Wines (Clare Valley) Stonehaven (Padthaway) The Grapes of Ross (Barossa) Trentham Estate (Murray Darling) Two Tails Wines (Northern Rivers Zone) Villa d'Esta Vineyard (Northern Rivers Zone) Warraroong Estate (Hunter)

SAGRANTINO

Wine Type: Very full bodied wines with high acidity.

References: C, D, H, O

Sagrantino is a red wine grape from Central Italy. It is the basis for the powerful Montefalco wines in Umbria. Better winemaking techniques are now producing wines with the tannins under control, which was not always the case. Sagrantino has the potential to play a supporting role in the constellation of emerging Italian stars in Australia.

- Andrew Peace Wines (Swan Hill) Chalmers (Murray Darling) D'Arenberg (McLaren Vale) Domain Day (Barossa) Gracebrook Vineyards (King Valley) Heathvale (Eden Valley) Rupert's Ridge Estate (Heathcote)

SAINT MACAIRE

Saint Macaire is an obscure Bordeaux red wine variety that has all but disappeared in France. It is represented in Australia by some small plantings in the Riverina and Coonawarra. Westend Estate is using the variety to make the medal winning Calabria Private Bin Saint Macaire.

- Hollick (Coonawarra) Westend Estate (Riverina)

SANGIOVESE

Synonyms: Brunello, Sangiovese Dolce, Sangiovese Gentile, Sangiovese di Lamole, Morinello, Prugnolo, Calabrese, Sangioveto

Maturity group: This variety ripens in midseason, Wine Type: Full bodied wines with high acidity.

References: C, D, H, K, O

The name Sangiovese looks like it comes from an Italian saint, but apparently it derives from Sanguis Jovis or 'Jupiter's Blood.' It is the major variety in the Tuscan region and thus Chianti wine, but is grown widely throughout Italy.

As a rustic everyday variety Sangiovese was plagued by a multitude of subvarieties and clones of varying quality. Wines made from the variety could range from very ordinary to excellent.

Since the 1970s the rules have been extended to allow a blending of up 10 percent of other varieties. It was soon discovered that the use of Cabernet Sauvignon as the minor component with Sangiovese leads to outstanding results, the so-called super Tuscans.

An attempt to introduce the variety into Western Australia was frustrated when it was discovered that the vines which were planted under the name Sangiovese were in fact Carnelian.

With well over 250 Australian wineries now using the variety, Sangiovese is leading the charge to Italian reds. Its success has inspired interest in a dozen or so other Italian varieties.

In Australia Sangiovese is most often made as a straight varietal, but blends with other Italian varieties and with Shiraz can also be found. Some of the descriptors used to describe the flavours of Sangiovese include cherries, violets, tomatoes and herbs.

- 919 Wines (Riverland) Aldinga Bay (McLaren Vale) Amadio (Adelaide Hills) Amulet Vineyard (Beechworth) Anderson Winery (Rutherglen) Andrew Peace Wines (Swan Hill) Angullong Wines (Orange) Avalon Wines (King Valley) Aventine Wines (Granite

Belt) Avonmore Estate (Bendigo) Barrecas (Geographe) Barton Creek (Central Ranges Zone) Barton Estate (Canberra) Belalie Bend (Southern Flinders Region) Belgrave Park Winery (South Coast Zone) Benwarin Wines (Hunter) Bimbadgen Estate (Hunter) Blackford Stable Wines (Adelaide Hills) Blue Metal Vineyard (Southern Highlands) Boggy Creek Vineyards (King Valley) Bottin Wines (McLaren Vale) Boyntons Feathertop (Alpine Valleys) Brindabella Hills (Canberra) Broken Gate Wines (Heathcote) Brokenwood Wines (Hunter) Brown Brothers (King Valley) Burnbrae Winery (Mudgee) Cape Mentelle (Margaret River) Capel Vale (Geographe) Cardinam Estate (Clare Valley) Carlei Estate (Yarra Valley) Casella (Riverina) Castagna Vineyard (Beechworth) Catspaw Farm (Granite Belt) Ceravolo Premium Wines (Adelaide Plains) Chain of Ponds (Adelaide Hills) Chalk Hill Winery (McLaren Vale) Chalmers (Murray Darling) Chapel Hill (McLaren Vale) Charlatan Wines (McLaren Vale) Chrismont (King Valley) Ciccone Estate (King Valley) Clearview Estate Mudgee (Mudgee) Cofield Wines (Rutherglen) Colvin Wines (Hunter) Connor Park (Bendigo) Copper Bull (Yarra Valley) Coriole (McLaren Vale) Counterpoint Vineyard (Pyrenees) Crittenden at Dromana (Mornington Peninsula) Crooked River Wines (Shoalhaven Coast) Dal Zotto Estate (King Valley) Dalfaras (Nagambie Lakes) Deviation Road (Adelaide Hills) Di Lusso Estate (Mudgee) Domain Day (Barossa) Dominic Versace Wines (Adelaide Plains) Doonkuna Wines (Canberra) Drews Creek Wines (Hunter) Dromana Estate (Mornington Peninsula) Edwards and Chaffey (McLaren Vale) Eldredge (Clare Valley) Farrell Estate (Murray Darling) Fighting Gully Road (Beechworth) Flat View Vineyard (Clare Valley) Flying Duck Estate (King Valley) Flying Fish Cove (Margaret River) Flynn's Wines (Heathcote) Foster e Rocco (Bendigo) Four Winds Vineyard (Canberra) Galli Estate (Sunbury) Gemtree Vineyards (McLaren Vale) Geographe Wines (Geographe) Gin Gin Wines (Queensland Coastal) Glenguin (Hunter) Goombaronga Park (Alpine Valleys) Gracebrook Vineyards (King Valley) Greenstone Vineyard (Heathcote) Grove Estate Wines (Hilltops) Hamiltons Bluff (Cowra) Heartland Wines (Limestone Coast) Hollick Wines (Coonawarra) Hopwood Winery (Goulburn Valley) Hugh Hamilton (McLaren Vale) Indigo Wine Company (Beechworth) Innocent Bystander (Yarra Valley) Jinks Creek Winery (Gippsland) Joadja Vineyards (Southern Highlands) Jylland Vineyard (Central Western Australian Zone) Kangarilla Road (McLaren Vale) Katoa Wines (Heathcote) Kevin Sobels Wines (Hunter) King River Estate (King Valley) Kingsley Grove (South Burnett) Knots Wines (Heathcote) Koltz (McLaren Vale) La Cantina King Valley (King Valley) Lethbridge Wines (Geelong) Little Bridge (Canberra) Little Wine Company (Hunter) Lowe Family Wines (Mudgee) Macquarie Grove Vineyards (Western Plains) Mad Dog Wines (Barossa) MadFish Wines (Margaret River) Maglieri (McLaren Vale) Massoni (Pyrenees) Mayfield Vineyard (Orange) McCrae Mist Wines (Mornington Peninsula) McIvor Estate (Heathcote) Melross Estate (Pyrenees) Michael Unwin Wines (Grampians) Michelini (Alpine Valleys) Minnow Creek (McLaren Vale) Mitchell (Clare Valley) Moama Wines (Perricoota) Monichino Wines (Goulburn Valley) Montefalco Vineyard (Porongurup) Monument Vineyard (Central Ranges Zone) Mopoke Ridge Winery (Shoalhaven Coast) Moppity Vineyards (Hilltops) Mount Burrumboot Estate (Heathcote) Mount Langi Ghiran Vineyards (Grampians) Mount Vincent Estate (Hunter) Mulligan Wongara Vineyard (Cowra) Mulyan (Cowra) Neagles Rock Vineyards (Clare Valley) New Glory (Goulburn Valley) Nova Vita Wines (Adelaide Hills) Nugan Estate (King Valley) Oakvale (Hunter) Pasut Family Wines (Murray Darling) Paulmara Estate (Barossa) Penmara (Hunter) Perrini Estate (Adelaide Hills) Piako Vineyards (Murray Darling) Pialligo Estate (Canberra) Piggs Peake Winery

(Hunter) Pikes (Clare Valley) Pindarie Wines (Barossa) Piromit Wines (Riverina) Pizzini Wines (King Valley) Poet's Corner (Mudgee) Pokolbin Estate (Hunter) Politini (King Valley) Primo Estate (Adelaide Plains) Prince Hill Wines (Mudgee) Pycnantha Hill Estate (Clare Valley) Quealy (Mornington Peninsula) Racecourse Lane Wines (Hunter) Ravensworth Wines (Canberra) Remarkable View Winery (Southern Flinders Region) Rigel Wines (Mornington Peninsula) Rimfire Vineyards (Darling Downs) Ringer Reef Winery (Alpine Valleys) Riverbank Estate (Swan Valley) Roberts Estate (Murray Darling) Rojo Wines (Port Phillip Zone) Rookery Wines (Kangaroo Island) Ruane Winery (Southern Highlands) Rutherglen Estates (Rutherglen) Salisbury Winery (Murray Darling) Scaffidi Estate (Adelaide Hills) Sedona Estate (Upper Goulburn) Serenella Estate (Hunter) Settlers Ridge (Margaret River) Sevenoaks Wines (Hunter) Severn Brae Estate (Granite Belt) Shadowfax Vineyard and Winery (Geelong) Shays Flat Vineyard (Pyrenees) Sherwood Estate (Hastings River) Silverfox Wines (Perricoota) Skimstone (Mudgee) Southern Highland Wines (Southern Highlands) St Ignatius Vineyard (Pyrenees) Starvedog Lane (Adelaide Hills) Stella Bella (Margaret River) Stevens Brook Estate (Perricoota) Stone Bridge Estate (Manjimup) Stonehaven (Padthaway) Stumpy Gully (Mornington Peninsula) Sutton Grange Winery (Bendigo) Swings & Roundabouts (Margaret River) Tahbilk (Nagambie Lakes) Tall Poppy (Murray Darling) Tallis Wine Company (Goulburn Valley) Talunga (Adelaide Hills) Tannery Lane (Bendigo) Tar and Roses (Nagambie Lakes) Tatachilla (McLaren Vale) Tatehams Wines (Clare Valley) Tempus Two (Hunter) Tenafeate Creek Wines (Adelaide Plains) The Deanery Vineyards (Adelaide Hills) The Islander Estate Vineyards (Kangaroo Island) The Pawn Wine Company (Langhorne Creek) Tilbrook Estate (Adelaide Hills) Tintara (McLaren Vale) Tintilla Wines (Hunter) Tombstone Estate (Western Plains) Torzi Matthews (Eden Valley) Totino Wines (Adelaide Hills) Tower Estate (Hunter) Uleybury Wines (Adelaide Zone) Vale Creek Wines (Central Ranges Zone) Varrenti Wines (Grampians) Vasarelli (Currency Creek) Vinea Marson (Heathcote) Vintara (Rutherglen) Virgara Wines (Adelaide Plains) W Wine of Mudgee (Mudgee) Warrenmang Vineyard (Pyrenees) Watchbox Wines (Rutherglen) Whistling Eagle Wines (Heathcote) Windemere Wines (Granite Belt) Windowrie Estate (Cowra) Wise Wine (Margaret River) Wood Park (King Valley) Woolybud (Kangaroo Island) Yalumba Wine Company (Barossa) Yanmah Ridge (Manjimup) Yarra Yering (Yarra Valley) Yarrh Wines (Canberra) Zilzie Wines (Murray Darling) Zonte's Footstep (Langhorne Creek)

SAPERAVI

Wine Type: Full bodied wines with high acidity.

References: C, D, H, O

This red wine variety which is grown throughout the former Soviet Union, but the consensus now seems to be that its origin is in Georgia. During the Soviet era poor vineyard and winery practices masked this variety's great potential.

Saperavi is known to produce deeply colored wines, the name means 'dyer' in Russian. Wines produced from this variety are known to age well, although high tannins can be a problem if they are consumed too young.

- Blue Metal Vineyard (Southern Highlands) Boyntons Feathertop (Alpine Valleys) Domain Day (Barossa) Gapsted (Alpine Valleys) Hawkers Gate (McLaren Vale) Hugh Hamilton (McLaren Vale) Massena Wines (Barossa) Oak Works (Riverland) Ridgemill Estate (Granite Belt) Rookery Wines (Kangaroo Island) Symphonia (King Valley) Ten Miles East (Adelaide Hills)

SAUVIGNON BLANC

This mainstream variety is outside the scope of this book.

SAVAGNIN

Wine Type: Light bodied wine with high acidity.

References: C, D, O

Savagnin is a white wine variety grown in Eastern France to make a curious style of wine called *Vin Jaune* - yellow wine. The wine is made from late picked grapes which and allowed to develop a yeast film or flor. If you think this sounds like sherry you are right, but the difference is that it is much cooler and there is no solero system used in Vin Jaune. Savagnin is identical to the variety called Traminer, of which the distinctly aromatic Gewürztraminer is a mutation.

Until very recently Savagnin was not known to be used commercially in Australia. It is now known that some vineyards in Australia which thought they were planting Albarino were in fact planting Savagnin.

Savagnin was previously thought to be confined almost exclusively to the Jura region of Eastern France, a long way from Spain, Albarino's home. Bests at Great Western has a vine or two in its nursery block, previously thought to be the only specimens in a commercial vineyard.

Identifying grape varieties is not a straightforward task. There are clonal variations which make the job difficult. If the vines are grown in a different environment the vines make take on different characteristics and growth forms. Viruses are quite common in grape vines; they can have profound effects on the appearance of the vine.

The specialists who sort out all of this are called ampelographers. In the past they relied on systematic examination of the morphological characteristics of the vine, the leaves, the berries etc. In more recent years DNA profiling has added a more objective data, but ampelography is still a complex matter.

The wineries with Savagnin are now releasing the wine under its correct name. If you have some bottles from pre 2009 vintages labeled Albarino perhaps you should keep them as collectibles.

At best Australian Savagnins are light to medium bodied dry whites with persistence on the palate. Some great wines are being made even with a very short time in this country, and the variety certainly has a role in the varietal pantheon here.

- 919 Wines (Riverland) Allinda (Yarra Valley) Artwine (Clare Valley) Bago Vineyards (Hastings River) Banrock Station (Riverland) Brown Brothers (King Valley) Boyntons Feathertop (Alpine Valleys) Centennial Vineyards (Southern Highlands) Chapel Hill (McLaren Vale) Chrismont (King Valley) Crittenden at Dromana (Mornington Peninsula) Dunn's Creek Winery (Mornington Peninsula) Eumundi Winery (Queensland Coastal) First Drop (Barossa) Gemtree Vineyards (McLaren Vale) Glandore Estate (Hunter) Gracebrook Vineyards (King Valley) Hollick Wines (Coonawarra) Irvine (Eden Valley) Kellermeister Wines (Barossa) Mansfield Wines (Mudgee) Mosquito Hill Wines (Southern Fleurieu) Omersown Wines (Riverland) Plunkett Fowles (Strathbogie Ranges) Quarry Hill Wines (Canberra) Rusty Fig Wines (South Coast Zone) Stockman's Ridge (Central Ranges Zone) Symphonia (King Valley) Tamar Ridge (Northern Tasmania) Tscharke (Barossa) Tuck's Ridge (Mornington Peninsula)

SCHIOPPETTINO

Wine Type: Very full bodied wine with high acidity

References: C, D, O

Shioppettino is a red wine variety used in the North Eastern regions of Italy and in neighboring Slovenia.

Wines from this variety are full bodied with deep colour with aromatics redolent of Violets and pepper. It would probably do well in the cooler Australian wine regions but is yet to find a local champion among Australian winemakers.

- Chalmers (Murray Darling)

SCHONBURGER

Maturity group: Very Early maturing Wine Type: Full bodied wines with low acidity.

References: C, D, H, O

Schonburger is a German crossing with Pinot Noir as one of its parents. Hence it has pink berries although it is used for perfumed white wine. Its early maturity is reflected the distribution of Schonburger plantings, Germany, England, Canada and Tasmania.

In Australia interest in this variety seems to be confined to Tasmania.

- Barringwood Park (Northern Tasmania) Bream Creek Vineyard (Southern Tasmania) Freycinet (Southern Tasmania) Kraanwood (Southern Tasmania)

SEMILLON

This mainstream variety is outside the scope of this book.

SHALISTIN

This is a sport of Malian, which is itself a sport of Cabernet sauvignon. Shalistin therefore is a form of Cabernet sauvignon but without the pigments in the skin of the grapes. Shalastin produces a wine described as 'golden white'. The variety was developed on the Cleggett vineyard at Langhorne Creek who have registered the variety and make an attractive Shalistin dry white.

- Cleggett (Langhorne Creek)

SHIRAZ

This mainstream variety is outside the scope of this book.

SIEGERREBE

Maturity group: Very early maturity

References: H, K, O

An early ripening German white wine variety, noted for its ability to produce very high sugar levels. Jancis Robinson says it is grown, "like giant vegetables, purely by exhibitionists." The wines are noted for their strong aromatic character. Snowy Vineyard exploits the high sugar levels to make a white port style wine from Siegerrebe.

- Snowy Vineyard (Southern NSW) Wilmot Hills Vineyard (Northern Tasmania) Winery Palmara (Southern Tasmania)

SIRIUS

This is a recently bred hybrid grape with parents Bacchus (German Hybrid) and Villard (French Hybrid). The variety was bred in Germany with the aim of being able to produce well in a short growing season and to be resistant to Downy mildew. It is part of the program at Eumundi Vineyard to find varieties that will thrive in the climate which is conducive to fungal diseases.

- Eumundi (Queensland Coastal)

SOUZAO

Synonyms: Vinhao, Souson

Maturity group: Very late ripening variety

References: C, K, O

Souzao is a Portuguese red variety which is well regarded as a component of Port. Like other port varieties it can also be used to add a bit of fruit flavour and oomph to dry red blends. This is what D'Arenberg do when they make a wine called 'Sticks and Stones' a blend of Grenache, Tempranillo and Souzao.

- Burge Family Winemakers (Barossa) D'Arenberg (McLaren Vale) Mazza (Geographe) Peel Estate (Peel) Rivendell (Margaret River)

SYLVANER

Synonyms: Silvaner, Gruner silvaner, Franken, Osterreicher, Frankenriesling, Grunfrankish, Gamay Blanc, Gentil Vert, Johannisberger, Moravka

Maturity group: Early maturing, Group 3, Wine Type: Very light bodied wines with moderate acidity.

References: C, D, H, K, O

Also spelt Silvaner, this variety produces light dry white wines in Alsace, Germany, Switzerland and Italy's Alto Aldige. The variety is not very popular outside its native habitat probably because of the unfounded but widespread view that Sylvaner wines are devoid of character. Ballendeen Estate makes a well regarded late harvest Sylvaner.

- Angoves Winery (Riverland) Ballandeen Estate (Granite Belt) Felsberg Winery (Granite Belt) Hainault (Perth Hills) Home Hill (Southern Tasmania) James Estate (Hunter) Kellermeister Wines (Barossa) Mountilford (Mudgee) Snowy Vineyard (Southern New South Wales Zone) Stanley Brothers (Barossa)

TAMINGA

Maturity group: This variety ripens in midseason

References: H, K, O

Taminga is a white wine variety that was bred by the CSIRO in Australia for use as a high yielding variety in warm irrigation districts. Perhaps the most successful use of the variety is the Late Harvest Taminga produced by Trentham Estate.

- Clancy's of Conargo (Riverina) Eumundi Winery (Queensland Coastal) Hickinbotham (Mornington Peninsula) Kurrajong Downs (New England) Kyotmunga Estate (Perth Hills) Stakehill Estate (Peel) Trentham Estate (Murray Darling) Walla Wines (Big Rivers Zone)

TANNAT

Synonyms: Moustrou, Madiran, Harriague, Bordeleza

Maturity group: Late ripening variety, Wine Type: Very full bodied wines with high acidity.

References: C, D, H, O

Tannat's homeland is the south western area of France where one of its synonyms, Madiran, is also the name of an important appellation. Tannat is also popular in beef-eating Uruguay under the name of Harriague. Tannat wines are notable for their high levels of tannin. Careful winemaking or blending with Cabernet Sauvignon, Cab Franc or Fer can make these wines more approachable.

The level of health giving phenolics in Tannat based wines is believed to be responsible for longevity of people in the regions where it is the chief variety.

The variety is attracting some interest in Australia where it seems destined to be used for red wines intended for extended aging, or to lend its intense colour and astringency to dry red blends.

- Bago Vineyards (Hastings River) Boireann (Granite Belt) Bowe Lees (Adelaide Hills) Coolangatta Estate (Shoalhaven Coast) Deakin Estate (Murray Darling) Eumundi Winery (Queensland Coastal) Glenguin (Hunter) Goorambath (Glenrowan) Irvine (Eden Valley) Massena Wines (Barossa) Oak Works (Riverland) Pepper Tree Wines (Orange) Pertaringa (McLaren Vale) Pirramimma (McLaren Vale) Symphonia (King Valley) Symphony Hill Wines (Granite Belt) Toppers Mountain (New England) Trentham Estate (Murray Darling) Yacca Paddock Vineyards (Adelaide Hills) Zonte's Footstep (Langhorne Creek)

TARRANGO

Maturity group: Among the latest ripening varieties, Wine Type: Full bodied wines with low acidity.

References: C, D, K, O

Tarrango is red wine variety bred in Australia with the aim of producing wines with good colour and acidity, but low tannin. The aim of the CSIRO in breeding this wine was to have a variety suitable for hot irrigated areas and capable of making a lighter red style of wine.

- Brown Brothers (King Valley) Cynergie Wines (Goulburn Valley) Dookie College Winery (Goulburn Valley) Gapsted (Alpine Valleys) Rodericks (South Burnett) Stakehill Estate (Peel) Walla Wines (Big Rivers Zone)

TEMPRANILLO

Synonyms: Aragonez, Cencibel, Tinto Fino, Tinto Madrid, Grenache de Logrono, Tinto de la Rioja, Tinto de toro, Jacivera, Tempranilla, Ull de Llebre, Tinta Roriz, Tinto de Santiago

Maturity group: This variety ripens just before midseason. Wine Type: Full bodied wines with low acidity.

References: C, D, H, K, O

In Spain Tempranillo is the backbone of the wines of the Rioja and the Ribera del Duero region. In these regions it is often blended with varieties such as Graciano, Grenache or Cabernet sauvignon playing a minor role. It makes up about 60% of Ribera del Duero's famous Vega Sicilia, the Spanish equivalent to Grange. In Portugal the variety is used as a minor component in port, and some red table wines.

Elsewhere in the world the major plantings are in Argentina and California.

New plantings throughout Australian wine regions over the past decade have produced some very promising wines. The variety is versatile but perhaps does better in more continental areas.

What makes this variety so exciting? Well, it makes wines which have good colour and good fruit flavours along with low acid and low tannins. The middle palate is just full of interesting fruit flavours. This adds up to an easy drinking style of wine that matches very well with a range of foods.

The wine also goes well with American oak, a fact which many winemakers are exploring in Australia. In the vineyard the variety has a short growing season which makes it suitable for cooler areas.

82 VARIETIES • Tempranillo

Some Tempranillo wines are intended to be consumed while they are young, in which case they are referred to as being the joven style. The Spanish term crianza refers to aged wines. In very good years Spanish wineries make a reserva intended for extended aging.

- 919 Wines (Riverland) Amadio (Adelaide Hills) Anderson Winery (Rutherglen) Angelicus (Pemberton) Angoves Winery (Riverland) Artemis Wines (Southern Highlands) Artwine (Clare Valley) Audrey Wilkinson (Hunter) Avalon Wines (King Valley) Back Pocket (Granite Belt) Banrock Station (Riverland) Bella Ridge Estate (Swan District) Bidgeebong Wines (Southern New South Wales Zone) Blackbilly (McLaren Vale) Blue Poles Vineyard (Margaret River) Bogie Man Wines (Strathbogie Ranges) Boora Estate (Western Plains) Bousaada (Southern Highlands) Box Stallion (Mornington Peninsula) Boyntons Feathertop (Alpine Valleys) Brandy Creek Wines (Gippsland) Broke Estate (Hunter) Brookhampton Estate (Geographe) Brown Brothers (King Valley) Bullers Calliope (Rutherglen) Byrne and Smith (McLaren Vale) Capel Vale (Geographe) Capital Wines (Canberra) Carlaminda Estate (Geographe) Cascabel (McLaren Vale) Casella (Riverina) Centennial Vineyards (Southern Highlands) Ceres Bridge Estate (Geelong) Chalmers (Murray Darling) Chapel Hill (McLaren Vale) Charlies Estate Wines (Swan Valley) Chrismont (King Valley) Coolangatta Estate (Shoalhaven Coast) Counterpoint Vineyard (Pyrenees) Cow Hill (Beechworth) Crabtree of Watervale (Clare Valley) Crittenden at Dromana (Mornington Peninsula) D'Arenberg (McLaren Vale) David Treager (Nagambie Lakes) Dawson and Wills (Strathbogie Ranges) Deakin Estate (Murray Darling) Dimbulla (Hunter) Doctors Nose Wines (New England) Dogrock Winery (Pyrenees) Donnybrook Estate (Geographe) Donovan Wines (Grampians) Dos Rios (Swan Hill) Dumaresq Valley Vineyard (New England) Dunn's Creek Winery (Mornington Peninsula) Eaglerange Wines (Alpine Valleys) Edwards and Chaffey (McLaren Vale) Epsilon (Barossa) Eumundi Winery (Queensland Coastal) Evelyn County Estate (Yarra Valley) Ferguson Falls Winery (Geographe) Fighting Gully Road (Beechworth) Fox Gordon (Barossa) Freeman Vineyards (Hilltops) Galli Estate (Sunbury) Gapsted (Alpine Valleys) Gemtree Vineyards (McLaren Vale) Geographe Wines (Geographe) Gin Gin Wines (Queensland Coastal) Glandore Estate (Hunter) Glaymond Wines (Barossa) Golden Grove Estate (Granite Belt) Goundrey (Mount Barker) Gowrie Mountain Estate (Darling Downs) Granite Hills (Macedon Ranges) Granite Ridge Wines (Granite Belt) Greenstone Vineyard (Heathcote) Happs (Margaret River) Hare's Chase (Barossa) Hastwell and Lightfoot (McLaren Vale) Hay Shed Hill Wines (Margaret River) Heathcote II (Heathcote) Heathcote Winery (Heathcote) Hewitson (Barossa) Hidden Creek (Granite Belt) Hollick Wines (Coonawarra) Honey Moon Vineyard (Adelaide Hills) Hotham Ridge Winery (Peel) Houghton (Swan Valley) Hugh Hamilton (McLaren Vale) Inghams Skilly Ridge Wines (Clare Valley) Jester Hill Wines (Granite Belt) Kaesler (Barossa) Kellermeister Wines (Barossa) Kennedy (Heathcote) Kingston Estate (Riverland) Kinloch Wines (Upper Goulburn) Kirrihill (Adelaide Hills) Krinklewood (Hunter) Kulkunbulla (Hunter) La Linea (Adelaide Hills) Liebichwein (Barossa) Lillypilly Estate (Riverina) Little Wine Company (Hunter) Llangibby Estate (Adelaide Hills) Lucas Estate (Granite Belt) Macketh House Historic Vineyard (Pyrenees) Macquarie Grove Vineyards (Western Plains) MadFish Wines (Margaret River) Mallee Estate (Riverland) Mansfield Wines (Mudgee) Manton's Creek Vineyard (Mornington Peninsula) Mayford Wines (Alpine Valleys) Mazza (Geographe)

McCuskers Vineyard (Perth Hills) McHenry Hohnen (Margaret River) Melville Hill Estate Wines (New England) Morrisons of Glenrowan (Glenrowan) Moss Brothers (Margaret River) Mount Avoca (Pyrenees) Mount Burrumboot Estate (Heathcote) Mount Charlie Winery (Macedon Ranges) Mount Majura (Canberra) Mr Riggs Wine Company (McLaren Vale) Mt Samaria Vineyard (Goulburn Valley) Mundrakoona Estate (Southern Highlands) Murdock (Barossa) Nashwauk (McLaren Vale) Nepenthe (Adelaide Hills) Nugan Estate (King Valley) Olivers Taranga (McLaren Vale) Optimiste (Mudgee) Paxton (McLaren Vale) Peel Ridge (Peel) Peter Lehmann (Barossa) Piako Vineyards (Murray Darling) Pikes (Clare Valley) Pindarie Wines (Barossa) Plan B (Margaret River) Pokolbin Estate (Hunter) Pondalowie (Bendigo) Quarry Hill Wines (Canberra) Quattro Mano (Barossa) Quoin Hill (Pyrenees) Red Earth Estate (Western Plains) Red Edge (Heathcote) Remarkable View Winery (Southern Flinders Region) Ridgemill Estate (Granite Belt) Riverbank Estate (Swan Valley) Romantic Vineyard (Pyrenees) Rookery Wines (Kangaroo Island) Ross Estate Wines (Barossa) Rusty Fig Wines (South Coast Zone) Samuels Gorge (McLaren Vale) Sanguine Estate (Heathcote) Serafino Wines (McLaren Vale) Seville Hill (Yarra Valley) Smallfry Wines (Barossa) Stanton and Killeen Wines (Rutherglen) Starvedog Lane (Adelaide Hills) Stella Bella (Margaret River) Stockman's Ridge (Central Ranges Zone) Stuart Wines (Heathcote) Summit Estate (Granite Belt) Sutherland Estate (Yarra Valley) Swings & Roundabouts (Margaret River) Symphonia (King Valley) Symphony Hill Wines (Granite Belt) Tahbilk (Nagambie Lakes) Talijancich (Swan Valley) Talunga (Adelaide Hills) Tar and Roses (Nagambie Lakes) Te-Aro (Barossa) Tempus Two (Hunter) Ten Minutes by Tractor (Mornington Peninsula) Terrel Estate Wines (Riverina) The Grove Vineyard (Margaret River) The Pawn Wine Company (Langhorne Creek) Thomson Estate (Riverland) Tim Adams (Clare Valley) Tintara (McLaren Vale) Tobin Wines (Granite Belt) Tokar Estate (Yarra Valley) Toppers Mountain (New England) Trahna Rutherglen Wines (Rutherglen) Tscharke (Barossa) Two Dragons Wine (Currency Creek) Vale Vineyard (Mornington Peninsula) Vinaceous (Various) Vinifera Wines (Mudgee) Vintara (Rutherglen) Wallington Wines (Cowra) Walter Clappis Wine Co (McLaren Vale) Watchbox Wines (Rutherglen) Welshmans Reef Vineyard (Bendigo) West Cape Howe Wines (Denmark) Westend Estate (Riverina) Whale Coast Wines (Southern Fleurieu) Willow Bridge Estate (Geographe) Wilson Vineyard (Clare Valley) Windowrie Estate (Cowra) Windsors Edge (Hunter) Witchmount Estate (Sunbury) Wovenfield (Geographe) Wrattonbully Vineyards (Wrattonbully) Yacca Paddock Vineyards (Adelaide Hills) Yalumba Wine Company (Barossa) Zilzie Wines (Murray Darling) Zonte's Footstep (Langhorne Creek)

TEROLDEGO

Synonyms: Teroldigo, Tiroldola

Wine Type: Very full bodied wines with high acidity.

References: C, D, O

Teroldego is yet another little known Northern Italian wine grape variety. For decades in its native Trentino it has been producing good rustic wine with little attention. With care in the vineyard and the winery Teroldego can produce very good wine with firm tannins and great complexity.

- Michilini (Alpine Valleys) Zonte's Footstep (Langhorne Creek)

TINTA AMARELA

Synonyms: Portugal, Trincadeira Preta, Espadeira

Maturity group: Early maturing. Wine Type: Full bodied wines with low acidity.

References: C, D, K, O

Tinta Amarela is a Portuguese variety used for port in the Duoro Valley and table wines in Southern Portugal.

- Massena Wines (Barossa) Peel Estate (Peel) Smallfry Wines (Barossa)

TINTA NEGRA MOLE

Synonyms: Negra Mole, Negramoll

Wine Type: Full bodied wines with high acidity.

References: C, D, O

This variety is now the mainstay of the fortified wine industry on Madiera, having displaced the more traditional varieties. It is also used to make table wines and is used for this purpose by Kangarilla Road.

- Kangarilla Road (McLaren Vale) Oak Works (Riverland) Settlement Wines (McLaren Vale)

TINTO CAO

References: C, O

The name means "red dog" in Portuguese, but why you would call a grape variety by that name is anyone's guess. Tinta Cao is a port variety from the Duoro Valley where it is valued for the spicy and floral flavours it adds to port.

Tinto cao is used in Australia as an ingredient in fortified wines and in table wines.

- 919 Wines (Riverland) Happs (Margaret River) Mansfield Wines (Mudgee) Mazza (Geographe) Peel Estate (Peel) Pennyweight Winery (Beechworth) Rivendell (Margaret River) Stanton and Killeen Wines (Rutherglen) Windsors Edge (Hunter)

TOCAI FRUILANO

Synonyms Tocai, Tocai italico, Sauvignonasse, Vert

Wine type: Light bodied wine with medium acidity.

References: C, D, O

The identity and naming of this white wine grape variety has only been sorted out in the past few years although it has been grown in the Friuli region of North East Italy for many years. It often makes everyday drinking wine but can be used to make wines with elegance and structure.

Kathleen Quealy who pioneered Pinot Gris in Australia is now the sole producer of Australian Tocai Fruilano.

- Quealy (Mornington Peninsula)

TOURIGA

Synonyms: Touriga Nacional, Mortagua

Maturity group: This variety ripens in midseason. Wine Type: Wines with super heavy body and with moderate acidity.

References: C, D, H, K, O

The correct name for this variety is Touriga Nacional. There is another variety Touriga Franca, but in Australia the term Touriga nearly always refers to Touriga Nacional.

It is the dominant variety of the fortified wine industry in Portugal, but it is also used extensively for dry red table wine as well. High levels of tannins in table wines made from Touriga are often softened by blending. The main use of the variety in Australia is for fortified wines, but an increasing number of Australian wineries are using it for red table wines, either as a varietal or in blends.

- 919 Wines (Riverland) Burge Family Winemakers (Barossa Valley) First Drop (Barossa Valley) Fyffe Field (North East Victoria) Gapsted (Alpine Valleys) Grey Sands (Northern Tasmania) Kaesler (Barossa Valley) Kangderaar Vineyard (Bendigo) Mansfield Wines (Mudgee) Mazza (Geographe) Myattsfield Vineyard and Winery (Perth Hills) Norse Wines (Queensland Coastal) Old Mill Estate (Langhorne Creek) Peel Estate (Peel) Pennyfield Wines (Riverland) Pennyweight Winery (Beechworth) Pondalowie (Bendigo) Quattro Mano (Barossa) Rimfire Vineyards (Darling Downs) Rivendell (Margaret River) SC Pannell (McLaren Vale) Sevenhill Wines (Clare Valley) St Hallett (Barossa Valley) Stanton and Killeen Wines (Rutherglen) Starvedog Lane (Adelaide Hills) Surveyor's Hill Winery (Canberra) Toppers Mountain (New England)

Wagga Wagga Winery (Riverina) Windsors Edge (Hunter Valley) Woodonga Hill (Hilltops)

TREBBIANO

Synonyms: Ugni blanc, Turbianno, Rossetto, Rossola, Spoletino, Santoro, Perugino, Rusciola, Bobiano, Buzzetto, Clairette Ronde, St-Emilion, Roussan, White Shiraz, White Hermitage

Maturity group: Very late ripening variety. Wine Type: Very light bodied wines with high acidity.

References: C, D, H, K, O

Trebbiano is a ubiquitous white variety in its native Italy where it is often used as blending material. In France the variety is called Ugni Blanc and is used in the production of wine for distillation into Cognac, Armagnac and fortifying spirit.

Trebbiano is grown in most wine countries, but it has a reputation for producing highly acid but neutral wines, and its use is being supplanted by more varieties which promise to make wines with more character.

The variety has been in decline in Australia but a few winemaker enthusiasts are taking care with Trebbiano to produce wines with some character.

- Amulet Vineyard (Beechworth) Bella Ridge Estate (Swan District) Berton Vineyards (Riverina) Calico Town (Rutherglen) Campbells Wines (Rutherglen) Chateau Dorrien (Barossa) Cobbitty Wines (South Coast Zone) Melange Wines (Riverina) Mudgee Wines (Mudgee) Spinifex (Barossa) Summerfield (Pyrenees) Taminick Cellars (Glenrowan) Tyrrells (Hunter) Waybourne (Geelong)

TROLLINGER

Synonyms: Blauer Trollinger, Schiava

Maturity group: This variety ripens just before midseason. Wine Type: Light bodied wines with low acidity.

References: C, D, O

This is a German variety but the name probably refers to Tyrol in Austria. It is used in Germany to produce red (or dark pink) sweetish wines. Hahndorf Hill in Australia uses the variety in a blend to make an excellent dry rose.

- Hahndorf Hill (Adelaide Hills)

TYRIAN

References: H

This new red wine variety was bred by Australia's CSIRO. It is a hybrid of Cabernet Sauvignon and the Spanish variety Sumoll. The aim of the breeding program was to produce a high quality variety which would thrive in warm, dry areas. Tyrian ripens later than Cabernet Sauvignon and the juice has higher acid, thus allowing winemakers to make a stable wine in warmer conditions. The wines produced have a bright hue. The name Tyrian comes from tyrian purple, a bright purple dye used in the Ancient Mediterranean.

- McWilliams (Riverina)

VERDELHO

Synonyms: Gouveio

Maturity group: Early maturing, Wine Type: Wines with moderate body and with high acidity.

References: C, D, H, K, O

A native of Portugal and the island of Madeira, this white wine variety is used mainly for the production of fortified wines (White Port and Madeira.) Over recent years it has also been used for table wines. There is an unrelated Spanish variety called Verdejo.

In Australia, Verdelho has been used to make white table wines especially in Western Australia and the Hunter Valley in NSW. It has become increasingly popular in many other districts. It is at the stage of challenging Sauvignon blanc as a preferred variety for those seeking an alternative to Chardonnay.

The attitude of many wine writers to this variety can be summed up in the following quote: In his book *Varietal Wines*, James Halliday says "The success of Verdelho as a table wine in Australia is an extraordinary phenomenon without any obvious explanation."

But consumers love it. To anyone who puts their nose over a glass of Verdelho the explanation is very obvious. Verdelho is becoming more popular because the wines are straight-forward and they smell and taste like they are made from grapes. They will never be great wines, but many are just the ticket for informal lunches.

Some of the great 'Classic Dry Whites' of WA have a dollop of Verdelho in them along with Semillon, Sauvignon blanc and Chenin blanc.

88 VARIETIES • Verdelho

Verdelho is a suitable variety for warmer Australian regions and seems to handle humid ripening conditions better than some varieties.

- Alderley Creek Wines Estate (Northern Rivers Zone) Aldinga Bay (McLaren Vale) Allandale (Hunter) Ambar Hill (Granite Belt) Ambrook Wines (Swan Valley) Andrew Harris Vineyards (Mudgee) Angoves Winery (Riverland) Angullong Wines (Orange) Arimia Margaret River (Margaret River) Arrowfield (Hunter) Ashbrook Estate (Margaret River) Audrey Wilkinson (Hunter) Avalon Wines (King Valley) Baddaginnie Run (Strathbogie Ranges) Bago Vineyards (Hastings River) Bald Mountain (Granite Belt) Ballabourneen Wines (Hunter) Barambah Ridge Winery (South Burnett) Bawley Vale Estate (Shoalhaven Coast) Beckett's Flat (Margaret River) Beechwood Wines (Goulburn Valley) Beelgara Estate (Riverina) Belgenny Vineyard (Hunter) Bell's Lane Wines (Hunter) Bendigo Wine Estate (Bendigo) Bents Road (Granite Belt) Benwarin Wines (Hunter) Berrigan Wines (Swan District) Beyond Broke Vineyard (Hunter) Bidgeebong Wines (Southern New South Wales Zone) Big Hill Vineyard (Bendigo) Bimbadeen Estate (Hunter) Bimbadgen Estate (Hunter) Bishop Grove Wines (Hunter) Black George (Pemberton) Blackwood Wines (Blackwood Valley) Bleasdale (Langhorne Creek) Blue Manna (Margaret River) Blue Wren (Mudgee) Boatshed Vineyard (Hunter) Brammar Estate (Yarra Valley) Bremerton (Langhorne Creek) Briar Ridge Vineyard (Hunter) Bridgeman Downs Cellars (South Burnett) Briery Estate (Perth Hills) Broke's Promise (Hunter) Browns of Padthaway (Padthaway) Brush Box Vineyard (Hunter) Calais Estate (Hunter) Cambewarra Estate (Shoalhaven Coast) Camp Road Estate (Hunter) Camyr Allyn Wines (Hunter) Canonbah Bridge (Western Plains) Canungra Valley (Queensland Coastal) Cape Bouvard (Peel) Capel Vale (Geographe) Carosa (Perth Hills) Casella (Riverina) Cassegrain (Hastings River) Catherine's Ridge (Cowra) Catherine Vale Vineyard (Hunter) Cedar Creek Estate (Queensland Coastal) Centennial Vineyards (Southern Highlands) Channybearup (Pemberton) Chapel Hill (McLaren Vale) Chapman Valley Wines (Central Western Australian Zone) Chestnut Grove (Manjimup) Chidlows Well (Central Western Australian Zone) Chislehurst Estate (Hunter) Ciavarella (King Valley) Clancy's of Conargo (Riverina) Clovely Estate (South Burnett) Constable Vineyards (Hunter) Coolangatta Estate (Shoalhaven Coast) Cooper wines (Hunter) Crane Wines (South Burnett) Creeks Edge Wines (Mudgee) Crooked River Wines (Shoalhaven Coast) Dalfaras (Nagambie Lakes) David Hook Wines (Hunter) David Treager (Nagambie Lakes) De Beaurepaire Wines (Mudgee) De Bortoli (Riverina) De Iulius (Hunter) Deep Woods Estate (Margaret River) Divers Luck Wines (Northern Rivers Zone) Doctors Nose Wines (New England) Donegal Wines (Riverland) Donnybrook Estate (Geographe) Dos Rios (Swan Hill) Draytons (Hunter) Dusty Hill Estate (South Burnett) Echo Ridge Wines (Hunter) Elderton (Barossa) Elsmore's Caprera Grove (Hunter) Elysium Vineyard (Hunter) Emmas Cottage Vineyard (Hunter) Emmetts Crossing Wines (Peel) Ernest Hill Wines (Hunter) Eumundi Winery (Queensland Coastal) Evans and Tate (Margaret River) Faber Vineyard (Swan Valley) Fairview Wines (Hunter) Farrells Limestone Creek (Hunter) Fermoy Estate (Margaret River) First Creek (Hunter) Fish Tail Wines (Swan Valley) Flinders Bay (Margaret River) Flynn's Wines (Heathcote) Foate's Ridge (Hunter) Fonthill Wine (McLaren Vale) Fonty's Pool Vineyards (Pemberton) Fordwich Estate (Hunter) Fox Creek Wines (McLaren Vale) Francois Jacquard (Perth Hills) Fyffe Field (North East Victoria) Gapsted (Alpine Valleys) Garbin Estate (Swan Valley) Gartelmann Hunter (Hunter) Gecko Valley (Queensland Coastal) Gentle Annie (Goulburn Valley) Gin Gin Wines (Queensland

Coastal) Glandore Estate (Hunter) Glendonbrook (Hunter) Goorambath (Glenrowan) Goundrey (Mount Barker) Governor's Choice Winery (Queensland Zone) Gowrie Mountain Estate (Darling Downs) Granite Ridge Wines (Granite Belt) Great Lakes Wines (Northern Rivers Zone) Hackersley (Geographe) Halina Brook (Central Western Australian Zone) Hankin Estate (Goulburn Valley) Happs (Margaret River) Harrington Glen Estate (Granite Belt) Harris Organic Wines (Swan Valley) Harris River Estate (Geographe) Heafod Glen Winery (Swan Valley) Heartland Vineyard (Hunter) Heartland Wines (Limestone Coast) Henry's Drive (Padthaway) Heritage Estate (Granite Belt) Heron Lake Estate (Margaret River) Hidden Creek (Granite Belt) Hills View (McLaren Vale) Hillside Estate (Hunter) Hope Estate (Hunter) Hopwood Winery (Goulburn Valley) Houghton (Swan Valley) Hudson's Peake Wines (Hunter) Hugh Hamilton (McLaren Vale) Hunting Lodge Estate (South Burnett) Idle Hands Wines (Hunter) Illalangi Wines (Riverland) Inneslake (Hastings River) Iron Gate Estate (Hunter) Ironbark Hill Estate (Hunter) Island Brook (Margaret River) Ivanhoe Wines (Hunter) James Estate (Hunter) Jane Brook Estate (Swan Valley) Jarrah Ridge Winery (Perth Hills) Jarrets of Orange (Orange) Jasper Valley (Shoalhaven Coast) Jester Hill Wines (Granite Belt) Jimbour Wines (Queensland Zone) Jingalla (Great Southern) John Kosovich Wines (Swan Valley) Juniper Estate (Margaret River) Juul Wines (Hunter) Jylland Vineyard (Central Western Australian Zone) Kalari Wines (Cowra) Kancoona Valley Wines (Alpine Valleys) Karri Grove Estate (Margaret River) Keith Tulloch Wine (Hunter) Kevin Sobels Wines (Hunter) King River Estate (King Valley) Kingsley Grove (South Burnett) Kingston Estate (Riverland) Kladis Estate (Shoalhaven Coast) Knotting Hill Vineyard (Margaret River) Kooroomba (Queensland Zone) Krinklewood (Hunter) Kulkunbulla (Hunter) Lake Charlotte Wines (Perth Hills) Lamonts (Central Western Australian Zone) Lancaster Wines (Swan Valley) Laurellyn Wines (New England) Lawson Hill (Mudgee) Lighthouse Peak (Tumbarumba) Lilac Hill Estate (Swan Valley) Lilyvale Wines (Darling Downs) Lindenton Wines (Heathcote) Lindrum (Langhorne Creek) Little's Winery (Hunter) Little Wine Company (Hunter) Loch Luna (Riverland) Lost Valley (Upper Goulburn) Louee Wines (Mudgee) Lucas Estate (Granite Belt) Lucy's Run (Hunter) Mabrook Estate (Hunter) Madigan Vineyard (Hunter) Maleny Mountain Wines (Queensland Coastal) Margan Family (Hunter) Maroochy Springs (Queensland Coastal) Marybrook Vineyards & Winery (Margaret River) Mason Wines (Granite Belt) Maxwell Wines (McLaren Vale) McCuskers Vineyard (Perth Hills) McLeish Estate (Hunter) McPherson Wines (Nagambie Lakes) McWilliams (Riverina) Meera Park (Hunter) Melange Wines (Riverina) Melville Hill Estate Wines (New England) Middlesex 31 (Great Southern) Moama Wines (Perricoota) Moffatdale Ridge (South Burnett) Molly's Cradle (Hunter) Molly Morgan Vineyard (Hunter) Mollydooker Wines (South Australia) Monahan Estate (Hunter) Monument Vineyard (Central Ranges Zone) Moondah Brook (Swan Valley) Mopoke Ridge Winery (Shoalhaven Coast) Moss Brothers (Margaret River) Mount Appallan Vineyards (South Burnett) Mount Broke Wines (Hunter) Mount Burrumboot Estate (Heathcote) Mount Tamborine (Queensland Coastal) Mount View Estate (Hunter) Mount Vincent Estate (Hunter) Mudgee Growers (Mudgee) Murdup Wines (Mount Benson) Murrumbateman Winery (Canberra) Myattsfield Vineyard and Winery (Perth Hills) Neilson Estate Wines (Swan Valley) New Glory (Goulburn Valley) Nightingale Wines (Hunter) Normanby Wines (Queensland Zone) Norse Wines (Queensland Coastal) Nowra Hill Vineyard (Shoalhaven Coast) Nyora Vineyard and Winery (Gippsland) Oakover Estate (Swan Valley) Oakvale (Hunter) Olive Farm (Swan District) Olsen (Margaret River) Outram Estate (Hunter) Palmers Wines (Hunter) Peel

90 VARIETIES • Verdelho

Estate (Peel) Peel Ridge (Peel) Penmara (Hunter) Penny's Hill (McLaren Vale) Peos Estate (Manjimup) Peppin Ridge (Upper Goulburn) Pertaringa (McLaren Vale) Petersons Glenesk Estate (Mudgee) Pieter van Gent (Mudgee) Piggs Peake Winery (Hunter) Pinelli (Swan Valley) Piromit Wines (Riverina) Pokolbin Estate (Hunter) Polin & Polin (Hunter) Pothana (Hunter) Preston Peak (Granite Belt) Prince Hill Wines (Mudgee) Pyramid Hill Wines (Hunter) Pyramids Road Wines (Granite Belt) Racecourse Lane Wines (Hunter) Rangemore Estate (Darling Downs) Ravens Croft Wines (Granite Belt) Red Earth Estate (Western Plains) Red Tail (Northern Rivers Zone) Redgate (Margaret River) Reg Drayton (Hunter) Regent Wines (Swan Hill) Richfield Estate (New England) Ridgemill Estate (Granite Belt) Ridgeview Wines (Hunter) Rimfire Vineyards (Darling Downs) Rivendell (Margaret River) Riverbank Estate (Swan Valley) Riverina Estate Wines (Riverina) Robert Channon (Granite Belt) Roberts Estate (Murray Darling) Robyn Drayton (Hunter) Rodericks (South Burnett) Rosebrook Estate (Hunter) Roselea Estate (Shoalhaven Coast) Rothbury Ridge (Hunter) Rumbarella (Granite Belt) Rusty Fig Wines (South Coast Zone) Saddlers Creek Wines (Hunter) Salitage (Pemberton) Sam Miranda Wines (King Valley) Samson Hill Estate (Yarra Valley) Sandalford Wines (Swan Valley) Sandalyn Wilderness Estate (Hunter) Sarabah Estate (Queensland Zone) Serenella Estate (Hunter) Settlement Wines (McLaren Vale) Settlers Rise Montville (Queensland Coastal) Seven Mile Vineyard (Shoalhaven Coast) Sevenhill Wines (Clare Valley) Severn Brae Estate (Granite Belt) Shawwood Estate (Mudgee) Sherwood Estate (Hastings River) Silverfox Wines (Perricoota) Sirromet (Queensland Coastal) Sittella (Swan Valley) SmithLeigh Vineyard (Hunter) Southern Grand Estate (Hunter) Splitrock Vineyard Estate (Hunter) Springbrook Mountain Vineyard (Queensland Coastal) Stanton Estate (Queensland Zone) Stellar Ridge (Margaret River) Stevens Brook Estate (Perricoota) Stirling Wines (Hunter) Stomp (Hunter) Stringybark (Perth Hills) Stroud Valley Wines (Northern Rivers Zone) Stuart Range Estate (South Burnett) Summit Estate (Granite Belt) Sussanah Brook Wines (Swan District) Swanbrook Estate Wines (Swan Valley) Swooping Magpie (Margaret River) Symphony Hill Wines (Granite Belt) Tahbilk (Nagambie Lakes) Talijancich (Swan Valley) Tallavera Grove Winery (Hunter) Tamborine Estate Wines (Queensland Coastal) Tamburlaine (Hunter) Tangaratta Estate (Northern Slopes Zone) Tantemaggie (Pemberton) Tapestry (McLaren Vale) Tawonga Vineyard (Alpine Valleys) Temple Bruer (Langhorne Creek) Tempus Two (Hunter) The Grove Vineyard (Margaret River) The Natural Wine Company (Swan Valley) The Silos Estate (Shoalhaven Coast) Thomson Brook Wines (Geographe) Three Moon Creek (Queensland Zone) Tinonnee Vineyard (Hunter) Tipperary Estate (South Burnett) Tobin Wines (Granite Belt) Toppers Mountain (New England) Torambre Wines (Riverland) Tower Estate (Hunter) Tuart Ridge (Peel) Tulloch (Hunter) Twin Oaks (Queensland Coastal) Two Rivers (Hunter) Two Tails Wines (Northern Rivers Zone) Tyrrells (Hunter) Upper Reach Vineyard (Swan Valley) Vasse River Wines (Margaret River) Vercoes Vineyard (Hunter) Verona Vineyard (Hunter) Villa d'Esta Vineyard (Northern Rivers Zone) Vinaceous (Various) Wallambah Vale Wines (Northern Rivers Zone) Wandering Brook Estate (Peel) Wandin Valley Estate (Hunter) Wandoo Farm (Central Western Australian Zone) Waratah Vineyard (Queensland Zone) Warburn Estate (Riverina) Warraroong Estate (Hunter) Warrego (Queensland Zone) Wattle Ridge Wines (Blackwood Valley) Wattlebrook Vineyard (Hunter) Wells Parish Wines (Mudgee) Western Range Wines (Perth Hills) Westfield (Swan Valley) Whistle Stop Wines (South Burnett) White's Vineyard (Swan Valley) Wilkie Estate (Adelaide Plains) Willespie (Margaret River) Winbourne Wines (Hunter)

Windowrie Estate (Cowra) Windshaker Ridge (Perth Hills) Windsors Edge (Hunter) Windy Creek Estate (Swan Valley) Wirilda Creek (McLaren Vale) Wise Wine (Margaret River) Wonbah Estate (Queensland Coastal) Wonganella Wines (Northern Rivers Zone) Wood Park (King Valley) Woodstock (McLaren Vale) Woongoroo Estate (Queensland Coastal) Woop Woop Wines (McLaren Vale) Wordsworth Wines (Geographe) Wovenfield (Geographe) Yarrawa Estate (Shoalhaven Coast) Yarrawood (Yarra Valley) Yass Valley Wines (Canberra) Yokain Vineyard Estate (Geographe) Zonte's Footstep (Langhorne Creek)

VERDUZZO

Synonyms: Verduzzo Fruilano

References: C, H, O

This variety is used in North Eastern Italy to produce sweet and dry white wines. The sweet version is more highly regarded although it usually lacks the complexity of a great dessert wine.

- Bianchet (Yarra Valley) Lazzar Wines (Mornington Peninsula) Pepper Tree Wines (Orange) Pizzini Wines (King Valley) Vale Vineyard (Mornington Peninsula)

VERMENTINO

Synonyms: Vennentino, Malvoise, Pigato, Favorita

Wine Type: Light bodied wines with high acidity.

References: C, D, H, O

Vermentino is an aromatic white wine variety whose native habitat is the Italian region of Liguria and the Mediterranean islands of Corsica and Sardinia. It is becoming increasingly popular in Languedoc-Roussillon in France.

Vermentino has been successfully introduced into Australia over the past decade. It now has its own class at the Australian Alternative Varieties Wine Show.

Vermentino thrives in warmer climates producing well structured wines with pleasant aromatics. I have no doubt that this variety will play a major part in the future of the Australian wine industry.

- 919 Wines (Riverland) Aldinga Bay (McLaren Vale) Box Grove Vineyard (Nagambie Lakes) Boyntons Feathertop (Alpine Valleys) Brown Brothers (King Valley) Bullers Calliope (Rutherglen) Chalmers (Murray Darling) Cobaw Ridge (Macedon Ranges) Di Lusso Estate (Mudgee) Foxey's Hangout (Mornington Peninsula) Halifax (McLaren Vale) King River Estate)King Valley) Mansfield Wines (Mudgee) Parish Hill Wines (Adelaide Hills) Pasut Family Wines (Murray Darling) Politini (King Valley) Rimfire

Vineyards (Darling Downs) Rossiters (Murray Darling) Rupert's Ridge Estate (Heathcote) Spinifex (Barossa) Trentham Estate (Murray Darling) Vale Creek Wines (Central Ranges Zone) Yalumba Wine Company (Barossa) Zonte's Footstep (Langhorne Creek)

VILLARD BLANC

Synonyms: Seyve Villard

References: O

This variety is a French Hybrid, in other words a hybrid of the European vine species with an American species. Such hybrids are being eradicated in France where they are blamed for the production of the so called European wine lake. Villard Blanc is perhaps the most successful of the white French hybrids and is used in the Eastern United States.

In Australia the variety seems confined to the North Coast of NSW.

- Douglas Vale (Hastings River) Raleigh Wines (Northern Rivers Zone) Two Tails Wines (Northern Rivers Zone)

VIOGNIER

Synonyms: Vionnier

Maturity group: This variety ripens in midseason, Wine Type: Full bodied wines with low acidity.

References: C, D, H, K, O

The Viognier white grape variety was rescued from near extinction just a few decades ago and is now one of the hottest varietals going around.

Legend has it that the variety was introduced, along with Syrah, to the Rhone Valley during the Roman occupation. After two millennia of barbarian invasions, dark ages, wars and the ravages of time just a few hectares of Viognier were left by the 1960s. It occupied a small corner of the Rhone around Condrieu where it made cult wines known only to a small group of enthusiasts. But the word got out and thanks to modern viticultural techniques Viognier is now widely planted in the Languedoc and Ardeche in France as well as in Italy, California and Australia. Its future in the Rhone is now assured.

The ripening of the process of Viognier grapes is regarded as idiosyncratic. The fruit flavours seem to arrive in a rush at the end of ripening, so patience and a strong nerve is required to avoid picking too early. It is therefore quite likely that there will be strong variation from vintage to vintage.

VARIETIES • Viognier

Viognier is increasingly popular as the white partner in co pigmentation with Shiraz. Only a small proportion of the white variety is used, usually much less than 5%. The grapes are mixed prior to fermentation, in contrast to blending where the mixing takes place after fermentation and often some barrel aging. Co-pigmentation is claimed to increase the brightness and persistence of the colour of Shiraz and also to lift the flavour.

As a varietal white wine Viognier is often described using comparisons with the aromas of flowers, fruit and spice. They are best enjoyed with food, and are robust enough to be paired with quite aromatic or mildly spiced dishes. Viognier wines will probably age well, but they are very good young so why wait?

- Alan and Veitch (Adelaide Hills) Albert River (Queensland Coastal) Aldinga Bay (McLaren Vale) Alkoomi (Frankland River) Allies Wines (Mornington Peninsula) Allusion Wines (Southern Fleurieu) Amulet Vineyard (Beechworth) Anderson Winery (Rutherglen) Andrew Peace Wines (Swan Hill) Angoves Winery (Riverland) Angullong Wines (Orange) Arakoon (McLaren Vale) Artwine (Clare Valley) Arundel (Sunbury) Austin's Wines (Geelong) Avonmore Estate (Bendigo) Badgers Brook Yarra Valley (Yarra Valley) Bago Vineyards (Hastings River) Ballandean Estate (Granite Belt) Ballinaclash Wines (Hilltops) Balthazar (Barossa) Bartagunyah Estate (Southern Flinders Region) Barton Estate (Canberra) Barwick Wines (Margaret River) Battely Wines (Beechworth) Battle of Bosworth Wines (McLaren Vale) Battunga Vineyards (Adelaide Hills) Beelgara Estate (Riverina) Belgrave Park Winery (South Coast Zone) Belgravia Vineyards (Orange) Bellarine Estate (Geelong) Berton Vineyards (Riverina) Bimbadgen Estate (Hunter) Binbilla (Hilltops) Biscay Wines (Barossa) Blamires Butterfly Crossing (Bendigo) Blue Metal Vineyard (Southern Highlands) Blue Poles Vineyard (Margaret River) Boireann (Granite Belt) Box Grove Vineyard (Nagambie Lakes) Boyntons Feathertop (Alpine Valleys) Brammar Estate (Yarra Valley) Brave Goose Vineyard (Goulburn Valley) Brindabella Hills (Canberra) Brokenwood Wines (Hunter) Brookhampton Estate (Geographe) Brown Brothers (King Valley) By Farr (Geelong) Calais Estate (Hunter) Camp Road Estate (Hunter) Campbells Wines (Rutherglen) Cape Mentelle (Margaret River) Capel Vale (Geographe) Carickalinga Creek (Southern Fleurieu) Carilley Estate (Swan Valley) Carlaminda Estate (Geographe) Cascabel (McLaren Vale) Casella (Riverina) Castagna Vineyard (Beechworth) Ceres Bridge Estate (Geelong) Chain of Ponds (Adelaide Hills) Chalice Bridge Estate (Margaret River) Chalmers (Murray Darling) Charlies Estate Wines (Swan Valley) Chateau Mildura (Murray Darling) Ciavarella (King Valley) Circo V (King Valley) Claymore Wines (Clare Valley) Clonakilla (Canberra) Cobaw Ridge (Macedon Ranges) Conte Estate Wines (McLaren Vale) Coombe Farm Vineyard (Yarra Valley) Counterpoint Vineyard (Pyrenees) Cow Hill (Beechworth) Craneford (Barossa) Creed of Barossa (Barossa) Cupitt's Winery (Shoalhaven Coast) Currans Family Wines (Murray Darling) Cypress Post (Granite Belt) D'Arenberg (McLaren Vale) Darling Park (Mornington Peninsula) David Hook Wines (Hunter) David Treager (Nagambie Lakes) De Beaurepaire Wines (Mudgee) de Mestre Wines (Mudgee) Deakin Estate (Murray Darling) Diamond Valley Vineyards (Yarra Valley) Djinta Djinta (Gippsland) DogRidge (McLaren Vale) Domain Day (Barossa) Dos Rios (Swan Hill) Dowie Doole (McLaren Vale) Dyson Wines (McLaren Vale) Eaglerange Wines (Alpine Valleys) Eden Hall

VARIETIES • Viognier

(Eden Valley) Elgee Park (Mornington Peninsula) Farrell Estate (Murray Darling) Fighting Gully Road (Beechworth) First Creek (Hunter) Flying Duck Estate (King Valley) Flynn's Wines (Heathcote) Foggo Wines (McLaren Vale) Fonty's Pool Vineyards (Pemberton) Fox Gordon (Barossa) Francois Jacquard (Perth Hills) Frankland Estate (Frankland River) Freeman Vineyards (Hilltops) Galli Estate (Sunbury) Galvanized Wine Group (McLaren Vale) Geddes Wines (McLaren Vale) Gelland Estate (Mudgee) Gemtree Vineyards (McLaren Vale) Geoff Merrill (McLaren Vale) Gherardi Wines (Margaret River) Ghost Riders Vineyard (Hunter) Giant Steps (Yarra Valley) Glandore Estate (Hunter) Grant Burge (Barossa) Grey Sands (Northern Tasmania) Grove Estate Wines (Hilltops) Gundowringla Wines (Alpine Valleys) Haan (Barossa) Hamiltons Bluff (Cowra) Happs (Margaret River) Harris River Estate (Geographe) Hartley Estate (Perth Hills) Haselgrove (McLaren Vale) Hastwell and Lightfoot (McLaren Vale) Haywards of Locksley (Strathbogie Ranges) Heafod Glen Winery (Swan Valley) Heartland Vineyard (Hunter) Heartland Wines (Limestone Coast) Heathcote Winery (Heathcote) Heggies Vineyard (Eden Valley) Heidenriech Estate (Barossa) Helen's Hill Estate (Yarra Valley) Henley Hill (Yarra Valley) Henschke (Eden Valley) Hently Farm Wines (Barossa) Higher Plane (Margaret River) Hobbs of Barossa Ranges (Barossa) Honey Moon Vineyard (Adelaide Hills) Hotham Ridge Winery (Peel) House of Certain Views (Hunter) Howard Vineyard (Adelaide Hills) Hugh Hamilton (McLaren Vale) Indigo Wine Company (Beechworth) Inkwell (McLaren Vale) Innocent Bystander (Yarra Valley) Izway Wines (Barossa) Jamabro Wines (Barossa) Jarrah Ridge Winery (Perth Hills) Jeir Creek (Canberra) Jimbour Wines (Queensland Zone) Kaesler (Barossa) Kalgan River Wines (Albany) Kalleske Wines (Barossa) Kamberra (Canberra) Kangarilla Road (McLaren Vale) Katoa Wines (Heathcote) Kay Bros Amery (McLaren Vale) Keith Tulloch Wine (Hunter) Kidman Coonawarra Wines (Coonawarra) Kilikanoon (Clare Valley) Killara Estate (Yarra Valley) Kindred Spirit Wines (Strathbogie Ranges) King River Estate (King Valley) Kingston Estate (Riverland) Kirrihill (Adelaide Hills) Knotting Hill Vineyard (Margaret River) Koltz (McLaren Vale) Kouark (Gippsland) Ladbroke Grove (Coonawarra) Lady Bay Winery (Southern Fleurieu) Lamonts (Central Western Australian Zone) Langanook Wines (Bendigo) Langmeil (Barossa) Lanzthomson Wines (Barossa) Lark Hill Winery (Canberra) Lashmar (Kangaroo Island) Lerida Estate (Canberra) Lethbridge Wines (Geelong) Lillian (Pemberton) Lilliput Wines (Rutherglen) Linda Domas Wines (McLaren Vale) Lindenton Wines (Heathcote) Little's Winery (Hunter) Little River Wines (Swan Valley) Little Wine Company (Hunter) Logan Wines (Mudgee) Longview Vineyard (Adelaide Hills) Louee Wines (Mudgee) M. Chapoutier Australia (Mount Benson) Macquarie Grove Vineyards (Western Plains) Maddens Rise (Yarra Valley) Margan Family (Hunter) Mary Byrnes Wines (Granite Belt) mas serrat (Yarra Valley) Mason Wines (Granite Belt) Maxwell Wines (McLaren Vale) McCuskers Vineyard (Perth Hills) McHenry Hohnen (Margaret River) McIvor Creek (Heathcote) McKellar Ridge (Canberra) McPherson Wines (Nagambie Lakes) Meera Park (Hunter) Merum (Pemberton) Metier Wines (Yarra Valley) Mihi Creek Vineyard (New England) Miles from Nowhere (Margaret River) Millbrook Winery (Perth Hills) Milldale Estate Vineyard (Hunter) Mitchelton (Nagambie Lakes) Moaning Frog (Margaret River) Moppity Vineyards (Hilltops) Morrisons of Glenrowan (Glenrowan) Mount Appallan Vineyards (South Burnett) Mount Avoca (Pyrenees) Mount Buffalo Vineyard (Alpine Valleys) Mount Burrumboot Estate (Heathcote) Mount Camel Ridge Estate (Heathcote) Mount Cole Wineworks (Grampians) Mount Pilot Estate (North East Victoria) Mount Surmon (Clare Valley) Mount Trio Vineyard (Porongurup) Mountadam

(Eden Valley) Mr Riggs Wine Company (McLaren Vale) Mulyan (Cowra) Munari (Heathcote) Mundoonen (Canberra) Murray Street Vineyard (Barossa) Myattsfield Vineyard and Winery (Perth Hills) Myrtle Vale Vineyard (Upper Goulburn) Nalbra Estate (Geelong) Nepenthe (Adelaide Hills) Neqtar Wines (Murray Darling) Noorilim Estate (Goulburn Valley) Normanby Wines (Queensland Zone) Nursery Ridge (Murray Darling) Oakover Estate (Swan Valley) Oatley Wines (Mudgee) Oceanview Estates (Queensland Coastal) Olivers Taranga (McLaren Vale) Optimiste (Mudgee) Orange Mountain (Orange) Organic Vignerons Australia (Riverland) Paracombe Wines (Adelaide Hills) Parri Estate (Southern Fleurieu) Peel Ridge (Peel) Peerick Vineyard (Pyrenees) Pelican's Landing Maritime Wines (Southern Fleurieu) Pennan (Margaret River) Penny's Hill (McLaren Vale) Pennyfield Wines (Riverland) Pepper Tree Wines (Orange) Pepperilly Estate Wines (Geographe) Petaluma (Adelaide Hills) Petersons Glenesk Estate (Mudgee) Pettavel (Geelong) Philip Shaw (Orange) Phoenix Estate (Clare Valley) Piano Gully (Manjimup) Pikes (Clare Valley) Pirramimma (McLaren Vale) Plan B (Margaret River) Plunkett Fowles (Strathbogie Ranges) Poachers Ridge Vineyards (Mount Barker) Pondalowie (Bendigo) Possums Vineyard (McLaren Vale) Pothana (Hunter) Prince of Orange (Orange) Printhie Wines (Orange) Protero (Adelaide Hills) Purple Hen Wines (Gippsland) Pyren Vineyard (Pyrenees) Racecourse Lane Wines (Hunter) Ralph Fowler Wines (Mount Benson) Rangemore Estate (Darling Downs) Ravensworth Wines (Canberra) Redbank Victoria (King Valley) Redheads Studio (McLaren Vale) Rees Miller Estate (Upper Goulburn) Regent Wines (Swan Hill) Ridgemill Estate (Granite Belt) Ridgeview Wines (Hunter) Riverbank Estate (Swan Valley) Robert Johnson Vineyards (Eden Valley) Roberts Estate (Murray Darling) Rocky Passes Wines (Upper Goulburn) Romavilla (Roma) Roundstone Winery (Yarra Valley) Rowanston on the Track (Macedon Ranges) Rudderless Wines (McLaren Vale) Rupert's Ridge Estate (Heathcote) Rutherglen Estates (Rutherglen) Salisbury Winery (Murray Darling) Saltram (Barossa) Sanguine Estate (Heathcote) Sarabah Estate (Queensland Zone) Sautjan Vineyards (Macedon Ranges) Scion Vineyard (Rutherglen) Seabrook Wines (Barossa) Seven Ochres (Margaret River) Sevenhill Wines (Clare Valley) Shadowfax Vineyard and Winery (Geelong) Shelmerdine (Heathcote) Sieber Road Wines (Barossa) Sirromet (Queensland Coastal) Smallfry Wines (Barossa) Smidge Wines (Langhorne Creek) Snobs Creek Wines (Upper Goulburn) Solstice Mount Torrens Vineyards (Adelaide Hills) Sons of Eden (Barossa) Sorby Adams (Eden Valley) Southern Highland Wines (Southern Highlands) Spence Wines (Geelong) Spinifex (Barossa) Spoehr Creek Wines (Adelaide Hills) St Leonards (Rutherglen) Stanton and Killeen Wines (Rutherglen) Steinborner Family Vineyards (Barossa) Stella Bella (Margaret River) Sticks (Yarra Valley) Stone Ridge (Granite Belt) Stonehaven (Padthaway) Stuart Wines (Heathcote) Sugarloaf Ridge (Southern Tasmania) Sutherlands Creek Vineyard (Geelong) Sutton Grange Winery (Bendigo) Swings & Roundabouts (Margaret River) Symphonia (King Valley) Symphony Hill Wines (Granite Belt) Syrahmi (Heathcote) T'Gallant (Mornington Peninsula) Tahbilk (Nagambie Lakes) Tall Poppy (Murray Darling) Tallarook Wines (Upper Goulburn) Tallis Wine Company (Goulburn Valley) Taltarni (Pyrenees) Tamar Ridge (Northern Tasmania) Tanglewood Vines (Blackwood Valley) Tapestry (McLaren Vale) Tatachilla (McLaren Vale) Tawonga Vineyard (Alpine Valleys) Taylors (Clare Valley) Telgherry (Hunter) Temple Bruer (Langhorne Creek) Terra Felix (Upper Goulburn) The Islander Estate Vineyards (Kangaroo Island) The Lane (Adelaide Hills) The Pawn Wine Company (Langhorne Creek) The Ritual (Peel) The Standish Wine Company (Barossa) Three Moon Creek (Queensland Zone) Tiltili

Wines (Langhorne Creek) Tim Smith Wines (Barossa) Torbreck Vintners (Barossa) Trahna Rutherglen Wines (Rutherglen) Trentham Estate (Murray Darling) Turners Crossing Vineyard (Bendigo) Two Dorks Estate (Heathcote) Tyrrells (Hunter) Valhalla Wines (Rutherglen) Veritas (Barossa) Vincognita (McLaren Vale) Vinea Marson (Heathcote) Vintara (Rutherglen) Voyager Estate (Margaret River) W Wine of Mudgee (Mudgee) Wallington Wines (Cowra) Walter Clappis Wine Co (McLaren Vale) Wandoo Farm (Central Western Australian Zone) Wanted Man (Heathcote) Waratah Vineyard (Queensland Zone) Warrego (Queensland Zone) Watershed Wines (Margaret River) Waurn Ponds Estate (Geelong) Wedgetail Ridge Estate (Darling Downs) West Cape Howe Wines (Denmark) Westend Estate (Riverina) Western Range Wines (Perth Hills) Westgate Vineyard (Grampians) Westlake Vineyards (Barossa) Whale Coast Wines (Southern Fleurieu) Whicher Ridge (Geographe) Whistling Eagle Wines (Heathcote) Whitsend Estate (Yarra Valley) Whyworry Wines (New England) Wildwood (Sunbury) Willow Bridge Estate (Geographe) Wills Domain Vineyard (Margaret River) Willunga 100 Wines (McLaren Vale) Winbirra Vineyard (Mornington Peninsula) Winewood (Granite Belt) Wirra Wirra (McLaren Vale) Wood Park (King Valley) Woop Woop Wines (McLaren Vale) Wovenfield (Geographe) Wrattonbully Vineyards (Wrattonbully) Yalumba Wine Company (Barossa) Yangarra Estate (McLaren Vale) Yarra Yarra (Yarra Valley) Yarra Yering (Yarra Valley) Yarraloch (Yarra Valley) Yengari Wine Company (Beechworth) Yering Station (Yarra Valley) Zilzie Wines (Murray Darling) Zonte's Footstep (Langhorne Creek)

ZIBIBBO

Synonyms: Muscat of Alexandria

References: C, O

This is the Italian name for Muscat of Alexandria variety. The name has been adopted to denote a style of light pink sparkling wine, a stablemate for the more popular Moscato.

- Brown Brothers (King Valley) Crabtree of Watervale (Clare Valley) Pangallo Estate (Hunter)

ZINFANDEL

Synonyms: Primitivo, Plavic Veliki,

Maturity group: This variety ripens just before midseason, Wine Type: Wines with super heavy body and with moderate acidity.

References: C, D, H, K, O

Debate and discussion about its origin has raged over the past few years, but it is now established that the Californian Zinfandel is genetically identical to the Italian variety Primitivo and the obscure Croatian variety Crljenak Kastelanski.

The last named has the oldest provenance so in some ways can be regarded as the correct name. How Zinfandel and Primitivo got to their current homes is still an open topic.

This variety is popular in California where it makes all manner of wines; perhaps the best are the rugged dry reds where finesse is not an issue. Most popular is 'white Zin' which is a sweetish pink style that gets short shrift from most serious wine critics.

Under the name of Primitivo this variety is grown in Puglia in Southern Italy where it makes powerful, if rustic, reds. After a period in decline it is making a comeback there as new producers are treating it with respect and using it to make varietal wines rather than using it to bulk out blends.

Zinfandel has a reputation for being difficult to manage in the vineyard, and it is not user friendly in the winery either, so it seems that wine producers either love it or hate it.

It is prone to uneven ripening even within individual bunches which presents dilemmas at vintage time. Some winemakers, such as Rick Glastonbury at Kabminye in the Barossa Valley, see this as an advantage, allowing wine to be made with complex palate showing a range of ripe and ripening fruit flavours.

In Australia, Cape Mentelle at Margaret River and Nepenthe in the Adelaide Hills were the first to establish a reputation for producing good Zinfandel wines, but they now have plenty of competitors. It is fair to say that wines from this variety in Australia are works in progress, but most producers seem to be aiming at a robust red style rather than the sometimes insipid "White Zin" beverage.

Zinfandel has been used in fortified wines, Mansfield Wines of Mudgee are using it as a late picked variety and it has also been used to make sparkling red.

- Arimia Margaret River (Margaret River) Audrey Wilkinson (Hunter) Baarrooka (Strathbogie Ranges) Barrecas (Geographe) Brown Brothers (King Valley) Buckshot Vineyard (Heathcote) Burge Family Winemakers (Barossa) Calais Estate (Hunter) Cape Horn Vineyard (Goulburn Valley) Cape Mentelle (Margaret River) Cargo Road Wines (Orange) Chapman Valley Wines (Central Western Australian Zone) Chateau Tanunda (Barossa) Chittering Valley Winery (Perth Hills) Concotton Creek (Peel) De Bortoli (Riverina) Donnybrook Estate (Geographe) Elderton (Barossa) Glaymond Wines (Barossa) Grove Estate Wines (Hilltops) Hanging Rock Winery (Macedon Ranges) Hently Farm Wines (Barossa) Hotham Ridge Winery (Peel) Inghams Skilly Ridge Wines (Clare Valley) Inkwell (McLaren Vale) Irvine (Eden Valley) Jarrah Ridge Winery (Perth Hills) jb Wines (Barossa) Kabminye Wines (Barossa) Kangarilla Road (McLaren Vale) Karanto Vineyards (Langhorne Creek) Kingston Estate (Riverland) Lethbridge Wines (Geelong) Lilac Hill Estate (Swan Valley) Lion Mill Winery (Perth Hills) Longview Vineyard (Adelaide Hills) Lowe Family Wines (Mudgee) Mandalay Road (Geographe) Mansfield Wines (Mudgee) Marschall Groom Cellars (Barossa)

VARIETIES • Zinfandel

McAdams Lane (Geelong) Morrisons of Glenrowan (Glenrowan) Murray Street Vineyard (Barossa) Nepenthe (Adelaide Hills) O'Regan Creek Vineyard and Winery (Queensland Coastal) Oak Works (Riverland) Old Caves (Granite Belt) Olssens of Watervale (Clare Valley) Peel Estate (Peel) Pepper Tree Wines (Orange) Petersons Glenesk Estate (Mudgee) Phillips Estate (Pemberton) Piggs Peake Winery (Hunter) Rigel Wines (Mornington Peninsula) Riverbank Estate (Swan Valley) Robinvale Wines (Murray Darling) Romavilla (Roma) Rusden Wines (Barossa) Rusticana (Langhorne Creek) Rutherglen Estates (Rutherglen) Sanguine Estate (Heathcote) Schulz Vignerons (Barossa) Seraph's Crossing (Clare Valley) Smallwater Estate (Geographe) Smidge Wines (Langhorne Creek) Stellar Ridge (Margaret River) Summit Estate (Granite Belt) Sutherlands Creek Vineyard (Geelong) Tempus Two (Hunter) Tim Adams (Clare Valley) Tscharke (Barossa) Uleybury Wines (Adelaide Zone) Vincognita (McLaren Vale) Virage (Margaret River) Wandering Lane (Peel) Wandoo Farm (Central Western Australian Zone) Watershed Wines (Margaret River) Wilson Vineyard (Clare Valley) Wise Wine (Margaret River) Wood Park (King Valley) Wordsworth Wines (Geographe) Yokain Vineyard Estate (Geographe) Zonte's Footstep (Langhorne Creek)

Wine regions

Australian wine regions are organized into a system called Geographic Indicators. The system has regions as its basis but there are also subregions, zones and superzones. The arrangements are messy because of the need for compromise and the structure of the industry which is scattered very unevenly over a continent with a multitude of climates and stages of development.

The wineries listed with each region description are using alternative varieties, details of which can be found in the wineries listing in the latter part of this book.

SOUTH AUSTRALIA

South Australia claims to be the Wine State of Australia, based on the fact that it produces the most wine, even though Victoria has more wineries.

ADELAIDE SUPER ZONE

Unlike the other states South Australia as a "Super Zone" This encompasses three zones, the Barossa Zone, Mount Lofty Ranges Zone, and the Fleurieu Zone and therefore all ten regions in those zones. The concept of a Super Zone was the device used to satisfy those winemakers who make wine from grapes grown in several zones but wish to indicate the origin of the wine more specifically than the more generic South Eastern Australia.

Just two wineries covered by this book exist in the Adelaide Super Zone, but not in one of its component zones or regions.

- Caught Redhanded, Uleybury Wines

BAROSSA ZONE

There are two regions in this Zone, the Barossa Valley and Eden Valley. Although these two regions share a boundary and many wineries draw from grapes made in both regions, they are quite distinct. The Eden Valley is 100-200 metres higher and hence cooler. This has quite significant effects on the varieties grown, vintage time, and the styles of wine produced.

Barossa Valley Region

One of the best known regions of Australia, the Barossa Valley is just an hour's drive north of Adelaide. Major towns include Nuriootpa, Tanunda, Lyndoch and Angaston. The climate is warm and dry and the region is famed for its red wines although a considerable amount of white wine is also made.

- 1847, B3 Wines, Balthazar, Barossa Valley Estate, Basedow, Bethany, Biscay Wines, Burge Family Winemakers, Charles Melton, Chateau Dorrien, Chateau Tanunda, Cirillo, Clancy Fuller, Colonial Estate, Craneford, Creed of Barossa, Deisen, Diggers Bluff, Domain Barossa, Domain Day, Elderton, Eperosa, Epsilon, First Drop, Fox Gordon, Gibson Barossavale, Glaetzer Wines, Glaymond Wines, Gnadenfrei Estate, Gomersal Wines, Gomersal Wines, Grant Burge, Greenock Creek Wines, Haan, Hamiltons Ewell Vineyards, Hare's Chase, Heidenriech Estate, Henry Holmes Wines, Hently Farm Wines, Hewitson, Hobbs of Barossa Ranges, Izway Wines, Jamabro Wines, jb Wines, Jenke Vineyards, John Duval Wines, Kabminye Wines, Kaesler, Kalleske Wines, Kellermeister Wines, Kies Family, Kurtz Family Vineyards, Landhaus Estate, Langmeil, Lanzthomson Wines, Laughing Jack, Liebichwein, Limb Vineyards, Linfield Road Wines, Loan Wines, Lou Miranda Estate, Mad Dog Wines, Magpie Estate, Mardia Wines, Marschall Groom Cellars, Massena Wines, Maverick Wines, Moppa Wilton Vineyards, Murdock, Murray Street Vineyard, Orlando, Paulmara Estate, Peter Lehmann, Pindarie Wines, Quattro Mano, RBJ, Rockford, Roehr, Roennfeldt Wines, Rosenvale Wines, Ross Estate Wines, Rusden Wines, Saltram, Schild Estate Wines, Schiller Vineyards, Schubert Estate, Schulz Vignerons, Schwarz Wine Company, Scorpiiion, Seabrook Wines, Sieber Road Wines, Smallfry Wines, Sons of Eden, Soul Growers, Spinifex, St Hallett, Stanley Brothers, Steinborner Family Vineyards, Stonewell Vineyards, Tait Wines, Te-Aro, Teusner, The Grapes of Ross, The Standish Wine Company, Thorn-Clarke Wines, Tim Smith Wines, Torbreck Vintners, Tscharke, Turkey Flat Vineyards, Veritas, Veronique, Westlake Vineyards, Winter Creek Wine, Yaldara, Yalumba Wine Company, Yelland and Papps, Zitta Wines

Eden Valley

This region abuts the Barossa Valley and differs in that it is higher and cooler. The Eden Valley has a more rugged topography and hence a wider range of microclimates. There are more white wines, especially Rieslings, made from grapes grown in the Eden Valley than the Barossa Valley.

- Eden Hall, Eden Road Wines, Fernfield Wines, Hartz Barn Wines, Heathvale, Heggies Vineyard, Henschke, Hutton Vale, Irvine, Mountadam, Pewsy Vale, Robert Johnson Vineyards, Sorby Adams, Tin Shed Wines, Torzi Matthews, Wroxton Wines

FAR NORTH ZONE

This zone covers the vast bulk of arid inland South Australia but there are only a few wineries and these are all in the zone's only region - the Southern Flinders Ranges Region.

Southern Flinders Ranges Region

This area is known for its scenic beauty rather than its vineyards, but there are a few. Much of the production is sold to wineries further south but there are a couple of wineries here. In this very warm region red wines predominate.

- Bartagunyah Estate, Belalie Bend, Remarkable View Winery

FLEURIEU ZONE

This Zone contains five regions: Currency Creek, Kangaroo Island, Langhorne Creek, McLaren Vale and Southern Fleurieu. All are influenced by proximity to the sea. There is only one winery covered by this book which is inside the Fleurieu Zone but outside all of the regions.

- Springs Hill Vineyard

Currency Creek Region

This region is on the western shores of Lake Alexandrina at the mouth of the Murray River and to the south is the Southern Ocean. Hindmarsh Island is included in the region. Currency Creek Wine Region is newer and less well known than its neighbor Langhorne Creek. Most of the wine produced here is red from Cabernet Sauvignon, Shiraz and Merlot.

- Ballast Stone Estate, Currency Creek, Salomon Estate, Two Dragons Wine, Vasarelli

Langhorne Creek

The wine industry dates back 150 years in this region. The area is a flat river delta with a benign climate for grape growing. A large proportion of the wines grown in the region are blended and sold by large wineries located elsewhere but the number of local wineries is growing. Red wine varieties dominate.

- Beach Road, Ben Potts Wines, Bleasdale, Bremerton, Casa Freschi, Cleggett, Gipsie Jack, Karanto Vineyards, Lake Breeze, Lindrum, Old Mill Estate, Rusticana, Smidge Wines, Temple Bruer, The Pawn Wine Company, Tiltili Wines, Wenzel Family Wines, Zonte's Footstep

Kangaroo Island

The wine industry is relatively young on this island which is a little surprising given the amount of limestone soil and the unequivocally maritime climate. The Island is also ideally placed to exploit the profitable synergy between tourism and wine. Mainly red wines are grown on the island.

- Hazyblur Wines, Lashmar, Rookery Wines, The Islander Estate Vineyards, Williams Springs Road, Woolybud

McLaren Vale Region

This region is located just south of Adelaide, a mixed blessing as demand for land for housing is pressuring the wine industry here. Vineyards are located on the plains next to the Gulf of St Vincent and into the foothills of the Southern Mt Lofty Ranges. Red wines dominate this region especially Shiraz, Cabernet Sauvignon and Grenache-Shiraz-Mourvedre (GSM) blends are most popular. Riesling makes excellent wines here but they are difficult to sell. Many

McLaren Vale winemakers are now turning to alternative varieties but the classic red varieties will dominate here for a long while yet.

- Aldinga Bay, Amicus, Arakoon, Battle of Bosworth Wines, Beechtree Wines, Bent Creek Vineyards, Blackbilly, Blown Away, Bottin Wines, Brini Estate, Byrne and Smith, Cape Barren Wines, Cascabel, Chalk Hill Winery, Chapel Hill, Charlatan Wines, Clarence Hill, Clarendon Hills, Classic McLaren Wines, Conte Estate Wines, Contessa Estate, Coriole, D'Arenberg, De Lisio Wines, Di Fabio Estate, Doc Adams, DogRidge, Dowie Doole, Dyson Wines, Edwards and Chaffey, Five Geese Hillgrove Wines, Foggo Wines, Fonthill Wine, Fox Creek Wines, Galvanized Wine Group, Geddes Wines, Gemtree Vineyards, Geoff Hardy, Geoff Merrill, Gilligan, Grancari Estate, Halifax, Haselgrove, Hastwell and Lightfoot, Hawkers Gate, Hills View, Hugh Hamilton, Inkwell, Kangarilla Road, Kay Bros Amery, Kimber Wines, Koltz, La Curio, Lazy Ballerina, Linda Domas Wines, Maglieri, Marienberg, Marius Wines, Maxwell Wines, McLaren Ridge Estate, McLaren Wines, Middlebrook Estate, Minnow Creek, Mr Riggs Wine Company, Nashwauk, Noon Winery, Oliverhill, Olivers Taranga, Paxton, Pende Valde, Penny's Hill, Pertaringa, Pirramimma, Possums Vineyard, Redheads Studio, Richard Hamilton Wines, Rudderless Wines, Samuels Gorge, SC Pannell, Scarpatoni Estate, Serafino Wines, Settlement Wines, Shingleback, Simon Hackett, Tapestry, Tatachilla, The Old Faithful Estate, Tintara, Twelve Staves Wine Company, Vincognita, Vinrock, Walter Clappis Wine Co, Willunga 100 Wines, Wirilda Creek, Wirra Wirra, Woodstock, Woop Woop Wines, Yangarra Estate

Southern Fleurieu Region

This region is the southerly continuation of McLaren Vale, but the terrain is much more undulating. There were vineyards in the region during the nineteenth century but the modern industry is only a couple of decades old. The maritime influence on the climate leads to the production of elegant wines, mainly reds.

- Allusion Wines, Carickalinga Creek, Lady Bay Winery, Minko, Mosquito Hill Wines, Mount Trafford, Mt Billy, Parri Estate, Pelican's Landing Maritime Wines, Whale Coast Wines

LIMESTONE COAST ZONE

This Zone occupies the (relatively) cool, wet south-east corner of South Australia. It includes the Coonawarra, Mt Benson, Padthaway and Wrattonbully Regions. Much of the region is dominated by a series of limestone ridges marking former coastlines.

The following wineries are located within the Zone but outside the outside the wine regions, or draw wine from several regions.

- Cape Banks, Heartland Wines, Herbert Vineyard, Port Robe

Coonawarra

This region was for a long time the preeminent Cabernet Sauvignon region in Australia, in more recent times it shares that honor with Margaret River. Excellent Shiraz and Merlot as well as blended reds are also made. Some beautiful whites made from Chardonnay and Riesling can also be found but they are regarded as a sideshow to the reds. Few alternative varieties are found here, there is little incentive to move away from the classics.

- Hollick Wines, Kidman Coonawarra Wines, Ladbroke Grove, Leconfield, Rymill Coonawarra

Mt Benson

This small region is located on a windy coastline between Robe and Kingston SE. The landscape is undulating with limestone outcrops through red soil (terra rossa). The vineyards are all new, the oldest being merely two decades old, but the wines show considerable promise. Again this region is mainly planted to classic red varieties.

- Kreglinger Estate, M. Chapoutier Australia, Murdup Wines, Ralph Fowler Wines

Padthaway

This region is a strip along the Riddoch Highway northwest of Naracoorte. It is slightly warmer and drier than Coonawarra to the south, but the soil types are similar. There are very large vineyards here which are mostly owned by large wineries elsewhere or grapegrowers who supply wineries outside the region. Only a small proportion of the wine is made locally and so the region has a very low visibility. Shiraz, Chardonnay and Cabernet Sauvignon are the predominant varieties grown.

- Browns of Padthaway, Henry's Drive, Stonehaven

Wrattonbully

This region is immediately to the north of Coonawarra and includes the regional town of Naracoorte. It has similar climate and soils to Coonawarra and can be regarded as an extension of that region, without the cache of the name.

- Koppamura Wines, Stone Coast Wines, Wrattonbully Vineyards

LOWER MURRAY ZONE

This Zone includes just one region, the Riverland.

Riverland Wine Region

This region stretches along the Murray River downstream from the border with Victoria. The vineyards are all irrigated because of the hot climate and high evaporation. It produces about a quarter of Australia's wine. Over the past decade there has been a shift in emphasis from quantity to quality as better varieties, including some alternative varieties, have been introduced.

- 919 Wines, Angoves Winery, Australian Old Vine Wines, Banrock Station, Bonneyview, Donegal Wines, Illalangi Wines, Kahlon Estate Wines, Kingston Estate, Loch Luna, Mallee Estate, Nelwood Wines, O'Donohue's Find, Oak Works, Omersown Wines, Organic Vignerons Australia, Pennyfield Wines, Salena Estate, Sigismondi Estate Wines, Spook Hill Wines, Thomson Estate, Torambre Wines

MOUNT LOFTY RANGES ZONE

This zone surrounds the City of Adelaide and it includes the Adelaide Hills Region to the southeast, the Adelaide plains region to the north and the Clare Valley further north. The regions are quite distinct and hence little can be written to summarize the overall terroir.

A few wineries in the zone lie outside the regions and are hence listed here.

- Macaw Creek Wine, Stone Bridge Wines

Adelaide Hills

The name of this region says it all. The hills are stretch north south from the Barossa and Eden valleys to McLaren Vale and Southern Fleurieu Regions. There are many distinct microclimates allowing a large number of varieties to be grown successfully. It is cool enough for the Pinots Noir and Gris. Adelaide Hills is the only region where white wines (narrowly) predominate over red, while the fame of other South Australian regions is much greater, some of the best wines come from the Adelaide Hills. There is an abundance of small wineries and winemakers who are pushing the varietal boundaries here.

- Abbey Rock Wines, Alan and Veitch, Alta Wines, Amadio, Arrivo, Ashton Hills, Barristers Block, Battunga Vineyards, BK Wines, Blackets, Blackford Stable Wines, Bockman, Bowe Lees, Chain of Ponds, Cloudbreak Wines, Deviation Road, Hahndorf Hill, Hillstowe, Honey Moon Vineyard, Howard Vineyard, Jupiter Creek Winery, Kenton Hill, Kirrihill, La Linea, Leabrook Estate, Llangibby Estate, Longview Vineyard, Mawson Ridge, Mylkappa Wines, Nepenthe, Nova Vita Wines, Paracombe Wines, Parish Hill Wines, Perrini Estate, Petaluma, Pike and Joyce, Protero, Scaffidi Estate, Sidewood Estate, Solstice Mount Torrens Vineyards, Spoehr Creek Wines, Starvedog Lane, Talunga, Ten Miles East, The Deanery Vineyards, The Lane, Tilbrook Estate, TK Wines, Tomich Hill, Totino Wines, Yacca Paddock Vineyards

Adelaide Plains

This is a hot dry region despite its proximity to the sea. It lies just to the north of the city. Sangiovese is relatively common here, no surprise given the Italian heritage of many of the winemakers.

- Ceravolo Premium Wines, Diloreto Wines, Dominic Versace Wines, Farosa Estate, Gawler River Grove, Old Plains, Primo Estate, Tenafeate Creek Wines, Versace Wines, Virgara Wines, Wilkie Estate

Clare Valley Region

The Clare Valley Wine Region is located just over a hundred kilometres north of Adelaide. While its name uses the word 'valley' the region is really a series of valleys and uplands with considerable variation in topography and microclimate.

This region is most famous for its crisp Rieslings and is now recognised, both in Australia and internationally as the premium region for dry wines from that variety. Shiraz and Cabernet Sauvignon are used to make impressive dry red wines here as well.

- Adelina Wines, Annie's Lane, Artwine, Australian Domaine Wines, Cardinam Estate, Claymore Wines, Crabtree of Watervale, Eldredge, Eyre Creek, Fireblock, Flat View Vineyard, Inghams Skilly Ridge Wines, Jeanneret Wines, Kilikanoon, Kirrihill Estates, Knappstein Wines, Mitchell, Mount Surmon, Neagles Rock Vineyards, Old Station, Olssens of Watervale, Phoenix Estate, Pikes, Pycnantha Hill Estate, Reilly's Wines, Robertson of Clare, Seraph's Crossing, Sevenhill Wines, Skillogalee, Tatehams Wines, Taylors, Tim Adams, Wendouree, Wilson Vineyard

THE PENINSULAS ZONE

This Zone includes just one Region The Southern Eyre Peninsula around the Port Lincoln area. As far as I know there are no alternative grape varieties used there, but the region has some potential.

VICTORIA

Victoria is the state with the largest number of wineries and they are spread right across the state. There is a wide variety of climates, from the hottest to the coldest in mainland Australia. Victoria is the heartland of small innovative grape growers and wineries.

CENTRAL VICTORIA ZONE

This zone contains five regions and a sub-region as listed below. The only winery outside the regions is the following

- Akrasi Wines

Bendigo Wine Region

Just 150km north-west of Melbourne this region has developed a solid reputation for red wines since its wine industry was revived in the 1970s. Shiraz is certainly the variety of choice closely followed by Cabernet Sauvignon. White varieties have been much less successful here and make up only a small proportion of the crush. The warm dry climate is ideal for red varieties and you would expect Sangiovese and especially Tempranillo to have a future here.

- Avonmore Estate, Bendigo Wine Estate, Big Hill Vineyard, Blamires Butterfly Crossing, Blanche Barkly, Chaperon Wines, Connor Park, Foster e Rocco, Glenwillow Vineyard, Harcourt Valley, Kangderaar Vineyard, Langanook Wines, Old Loddon Wines, Passing Clouds, Pondalowie, Pyramid Gold, Sandhurst Ridge, Sutton Grange Winery, Tannery Lane, Turners Crossing Vineyard, Waterwheel Wines, Welshmans Reef Vineyard, Yandoit Hill Winery

Goulburn Valley Wine Region

This region includes the Nagambie Lakes Sub Region. It is roughly triangular in shape with Seymour at its southern extremity and the Murray River between Yarrawonga and Echuca forming the Northern boundary. Viticulture has been continuous here since 1860. There is a slight preponderance of red wines over white. The warmer and dryer northern end depends on irrigation.

- Beechwood Wines, Brave Goose Vineyard, Broken River Vineyards, Cape Horn Vineyard, Cynergie Wines, Dookie College Winery, Gentle Annie, Hankin Estate, Heritage Farm, Hopwood Winery, Jaengenya Wines, Monichino Wines, Mt Samaria Vineyard, New Glory, Noorilim Estate, Silver Wings Winemaking, Tallis Wine Company

Nagambie Lakes Sub Region

This subregion is centred on the Township of Nagambie and the nearby historic Tahbilk Winery located on the banks of the Goulburn River. Tahbilk is best known for its role in keeping the Marsanne variety when everyone else was giving it away. The mild to warm climate and some deep sandy soils have allowed some vines to survive for decades and produce some powerful wines.

- Box Grove Vineyard, Dalfaras, David Treager, Goulburn Terrace, McPherson Wines, Mitchelton, Tahbilk, Tar and Roses, Twelve Acres

Heathcote Wine Region

Heathcote is another gold era town in Central Victoria whose wine industry faded away by the end of the nineteenth century to be revived in the last third of the twentieth. In recent years grape growers have been attracted to the deep friable soils which have developed on ancient rocks exposed by a geological fault.

Two wineries have achieved cult status for their shiraz based red wines, Jasper Hill and Wild Duck Creek Estate, indeed Heathcote Shiraz generally is well regarded by critics and consumers. However other wineries are using some Rhone varieties, as well as Italian varieties such as Sangiovese and Nebbiolo and more recently the Spanish Tempranillo.

- Armstead Estate, Broken Gate Wines, Buckshot Vineyard, Burke and Wills Winery, Flynn's Wines, Greenstone Vineyard, Heathcote Estate, Heathcote II, Heathcote Winery, Huntleigh Vineyards, Jasper Hill, Katoa Wines, Kennedy, Knots Wines, Lindenton Wines, Luke Lambert Wines, McIvor Creek, McIvor Estate, Mount Burrumboot Estate, Mount Camel Ridge Estate, Munari, Red Edge, Rogues Lane Vineyard, Rupert's Ridge Estate, Sanguine Estate, Shelmerdine, St Michael's Vineyard, Stuart Wines, Syrahmi, Toolleen Vineyard, Two Dorks Estate, Vinea Marson, Wanted Man, Whistling Eagle Wines

Strathbogie Ranges

The Strathbogie Ranges stretches along the eastern side of the Hume Highway from Seymour to Benalla. The landscape is rugged and largely unpopulated. There are only a few wineries, some larger operations outside the region have vineyards here. The higher vineyards are quite cold, suitable for Riesling as well as Chardonnay and Pinot Noir for sparkling wine production. Shiraz and Cabernet Sauvignon are used for red wine.

- Baarrooka, Baddaginnie Run, Bogie Man Wines, Dawson and Wills, Haywards of Locksley, Kindred Spirit Wines, Plunkett Fowles, Tarup Ridge Winery

Upper Goulburn Wine Region

This is another lesser known Victorian wine region, but the first winery in the region, Delatite, is fairly well known.

The region has an undoubtedly cool climate, with vines planted up to 800m in altitude. Red and white wine grapes are grown in almost equal numbers and the abundance of microclimates has encouraged growers to use a range of varieties.

- Delatite Winery, Glen Creek Wines, Growlers Gully, Kinloch Wines, Lost Valley, Melaleuca Grove, Myrtle Vale Vineyard, Peppin Ridge, Philip Lobe Wines, Rees Miller Estate, Rocky Passes Wines, Sedona Estate, Snobs Creek Wines, Tallarook Wines, Terra Felix, Tulley Wells

GIPPSLAND ZONE

Vineyards and wineries are so few and far between in Gippsland that there are no formal regions in this zone which stretches from the outskirts of Melbourne right along the coast to the NSW border. The climate is varied with the Australian Alps bordering the north west with Bass Strait to the South east. Among the multitude of the microclimates and potential sites for vineyards only a few are developed. Most sites are cool by Australian standards and the varietal selection reflects this with Pinot Noir, Cabernet Sauvignon, Chardonnay and Sauvignon blanc all prominent. The Zone is still in its pioneering phase and will continue to attract growers and winemakers who think outside the square.

- Ada River, Bass Phillip, Bass River, Brandy Creek Wines, Djinta Djinta, Jinks Creek Winery, Kongwak Hills Winery, Kouark, Lochmoore, Lyre Bird Hill, Moondarra, Mount Markey, Noorinbee Selection Vineyards, Nyora Vineyard and Winery, Purple Hen Wines, Sarsfield Estate, Silverwaters Vineyard, Tanjil Wines, Toms Cap, Wild Dog Winery, Windy Ridge Vineyard

NORTH EAST VICTORIA ZONE

This Zone contains five regions each with quite disparate viticultural histories, topographies, climates and wine styles. From a wine style point of view the Zone doesn't make much sense but the regions can be visited in a single trip.

A few wineries fall outside the declared regions in the Zone.

- Fyffe Field, Mount Pilot Estate, Murray Estate, Patrice Winery, Wirruna Estate

Alpine Valleys

This region covers the Kiewa, Ovens, Buffalo and Buckland valleys in the foothills to the Victorian snowfields. Varying altitudes allow both late and early ripening grapes to be grown. As is the case in the nearby King Valley the region was formerly a tobacco growing area which has made the switch to wine over the past few decades.

- Annapurna, Boyntons Feathertop, Eaglerange Wines, Folino Estate, Gapsted, Goombaronga Park, Gundowringla Wines, Kancoona Valley Wines, Mayford Wines, Michelini, Mount Buffalo Vineyard, Ringer Reef Winery, Souters Vineyard, Tawonga Vineyard

Beechworth

The hilly topography of this region means that most vineyards are small, but what the region lacks in quantity it certainly makes up for in quality. Two wineries established in the 1980s, Giaconda and Sorrenberg, have set a high standard and others are rising to the challenge of turning the excellent terroir into excellent wines. Red and white wines are produced here with potential for many varieties in the varied microclimates.

- Amulet Vineyard, Battely Wines, Bowmans Run, Castagna Vineyard, Cow Hill, Fighting Gully Road, Giaconda, Golden Ball, Indigo Wine Company, Pennyweight Winery, Sorrenberg, Yengari Wine Company

Glenrowan

This region can be considered as a south westerly extension of the Rutherglen; it shares a similar history, though not on such a grand scale, and it makes a similar range of wine. Many wineries are clustered along the western foothills of the Warby Range in sandy granitic soil. Red table wines and fortifieds predominate.

- Bailey's of Glenrowan, Goorambath, Judds Warby Range Estate, Morrisons of Glenrowan, Taminick Cellars

King Valley

This region has a warm, flat northern end grading to a cold mountainous south and the viticultural possibilities vary accordingly. There is a strong influence of the Italian heritage of many of the grapegrowers who have switched from suppliers of classic varieties to large wine companies to winemakers of an increasing range of alternative varieties. The King Valley is home to Brown Brothers the leading innovator in Australian wine varieties, but there are plenty of smaller wineries who are making their mark as well.

- Avalon Wines, Bergamin Wines, Brown Brothers, Boggy Creek Vineyards, Chrismont, Ciavarella, Ciccone Estate, Circo V, Collina del Re, Dal Zotto Estate, Darling Estate, Flying Duck Estate, Gracebrook Vineyards, Henderson Hardie, John Gehrig Wines, King River Estate, La Cantina King Valley, Nugan Estate, Paul Bettio, Pizzini Wines, Politini, Primerano, Reads, Redbank Victoria, Rose Hill Estate Wines, Sam Miranda Wines, Symphonia, Wood Park

Rutherglen

This region is the most distinctive and historically rich of all Victorian wine regions. Many of the wineries have been operated in one family for five or so generations, and the region's red brick buildings dating from the mid Nineteenth Century bear witness to the role Rutherglen has played in Australia's wine history. The hot days and cold nights characteristic of the summer and autumn allow for a range of varieties and styles to be successfully made.

Rutherglen's reputation was forged on fortified wine styles (once called Port, Muscat and Tokay) and big bodied tannic red table wines, based on Shiraz and other Rhone varieties. A more recent addition (since 1908) has been Durif red wines. By the end of the Twentieth Century new varieties were emerging including some interesting white wines, but the region is dominated by reds.

- All Saints Estate, Anderson Winery, Bullers Calliope, Calico Town, Campbells Wines, Chambers Rosewood, Cofield Wines, Drinkmoor Wines, Gehrig Estate, Jones Winery and Vineyard, Lake Moodemere, Lilliput Wines, Morris, Mount Prior, Pfeiffer Wines, Platt's, Rutherglen Estates, Scion Vineyard, St Leonards, Stanton and Killeen Wines, Trahna Rutherglen Wines, Valhalla Wines, Vintara, Warrabilla Wines, Watchbox Wines

NORTH WEST VICTORIA ZONE

This zone has two regions both of which cover part of New South Wales as well as Victoria, but the viticultural environment is similar on both sides of the Murray River so it makes sense to have the regions together.

Murray Darling

Australia's two great rivers converge just near Mildura, and much of the surrounding land has been used for various irrigated crops since the 1890s. The region produced fortified wines and brandy and since the 1930s it has been one of Australia's largest producers of table wines. Much of the wine is sold as bulk wine or in casks, but over the past couple of decades some attention has been give to producing quality table wines. Chalmers Vine Nursery in this region has done much to introduce new varieties and clones into Australia, and to encourage and educate winegrowers, winemakers and consumers to the possibilities of new wines.

- Callipari Wine, Carn Estate, Chalmers, Chateau Mildura, Currans Family Wines, Deakin Estate, Farrell Estate, Harkaway Estate, Mulcra Estate Wines, Neqtar Wines, Nursery Ridge, Pasut Family Wines, Piako Vineyards, Red Cliffs, Roberts Estate, Robinvale Wines, Rossiters, Salisbury Winery, Tall Poppy, Trentham Estate, Zilzie Wines

Swan Hill

This region produces far more wine than most people think. There are no iconic brands here, but many larger operations including Brown Brothers, have vineyards in the area. Red and fortified wines dominate but there is a significant amount of white wine produced as well.

- Andrew Peace Wines, Brumby Wines, Bullers Beverford, Carpenteri Vineyards, Date Brothers, Dos Rios, Oak Dale Wines, Regent Wines, Renewan Swan Hill, Yellymong

PORT PHILLIP ZONE

As the name implies the wine regions of this Zone surround Melbourne's Port Phillip Bay. All share the advantage of being within easy reach of a city of 4 million people may of whom are keen wine drinkers. A couple of wineries draw from grapes grown in various regions and are listed here.

- Chestnut Hill Vineyard, Rojo Wines

Geelong

The thriving nineteenth century wine industry in Geelong was utterly devastated by *Phylloxera*. The revival started in the 1960s but has not been as rapid as in other Victoria wine regions and has only gathered momentum since the turn of the new century. The Bellarine Peninsula enjoys a maritime influence on its climate while some of the more inland regions of the Geelong Wine Region have a drier and more varied climate.

Pinot Noir is the favoured variety in this region but a range of other varieties are also used. Pinot grigio and Riesling also do very well here.

- Amietta Vineyard, Austin's Wines, Baie Wines, Banks Road, Barwon Ridge Wines, Bellarine Estate, Bellbrae Estate, Brown Magpie Wines, By Farr, Ceres Bridge Estate, Clyde Park Vineyard, Curlewis Winery, Del Rios, Dinny Goonan Family Estate, Eagles Rise, Grassy Point Coatsworth Wines, Heytesbury Ridge, Jindalee Estate, Kilgour Estate, Kurabana, Lethbridge Wines, Leura Park Estate, McAdams Lane, Mount Anakie, Nalbra Estate, Peninsula Baie Wines, Pettavel, Pollocksford Vineyards, Provenance Wines, Scotchmans Hill, Shadowfax Vineyard and Winery, Spence Wines, Staunton Vale Vineyard, Sutherlands Creek Vineyard, Tarcoola Estate, The Minya Winery, Waurn Ponds Estate, Waybourne, Winchelsea Estate, Wyuna Park Vineyard

Macedon Ranges

The Macedon Ranges dominate the northern skyline of Melbourne and despite its proximity to the city some of the landscape still seems wild and remote. This is the coolest of Victoria's regions with plenty of Chardonnay and Pinot Noir grown for both sparkling and table wines. There are also impressive wines made from Shiraz, Cabernet sauvignon, and Riesling here. In more recent times Pinot gris has also been prominent.

- Basalt Ridge, Birthday Villa Vineyard, Cleveland, Cobaw Ridge, Curly Flat, Daylesford Wine Company, Ellender Estate, Fryerstown Road Vineyard, Glen Erin Vineyard Resort, Granite Hills, Hanging Rock Winery, Lancefield Winery, Lane's End Vineyard, Midhill Vineyard, MorganField, Mount Charlie Winery, Mount Franklin Estate, Mount Macedon Winery, Mount Towrong, Paramoor Wines, Pegeric, Portree Vineyard, Rowanston on the Track, Sailors Falls Winery, Sautjan Vineyards, Wili-Wilia Winery

Mornington Peninsula

In spite of its importance as a wine region many people are surprised to learn that viticulture has only been carried out on the Peninsula to any extent for the past four decades, most other Victorian regions had a nineteenth century antecedent. Urban development, recreational farming and seaside resorts have created an environment where small scale viticulture and wine tourism have contributed to a booming wine industry.

The cool maritime climate is unsuitable for Cabernet Sauvignon which was initially the favoured red variety here. Once this was realised Pinot noir became the variety of choice and now comprises the majority of red wines produced. Pinot gris was pioneered through the efforts of Kathleen Quealy, and the other great contributor to the development of the region is Garry Crittenden who provided the impetus for discovering how to grow grapes in the cool environment as well as championing Italian varieties like Arneis.

Allies Wines, Baillieu Vineyard, Barmah Park Wines, Barrymore Estate, Bayview Estate Winery, Bluestone Lane, Box Stallion, Crittenden at Dromana, Darling Park, Dromana Estate, Dunn's Creek Winery, Elan Vineyard, Eldridge Estate, Elgee Park, Ermes Estate, Five Sons Estate, Foxey's Hangout, French Island Vineyards, Harlow Park Estate, Hickinbotham, HPR Wines, Jones Road, Lazzar Wines, Lindenderry, Manton's Creek Vineyard, Maritime Estate, McCrae Mist Wines, Miceli, Montalto Vineyards, Moorooduc Estate, Morning Star Estate, Morning Sun Vineyard, Nazaaray, Ocean Eight Vineyard and Winery, Paradigm Hill, Paringa Estate, Phaedrus Estate, Pier 10, Point Leo Road Vineyard, Port Phillip Estate, Prancing Horse Estate, Quealy, Rahona Valley Vineyard, Red Hill Estate, Rigel Wines, Scorpo Wines, Seaforth Vineyard, Somerbury Estate, South Channel Wines, Stumpy Gully, T'Gallant, Tanglewood Downs, Ten Minutes by Tractor, The Cups Estate, The Garden Vineyard, Tuck's Ridge, Vale Vineyard, Vintina Estate, Whinstone Estate, Winbirra Vineyard, Yabby Lake Winery

Sunbury

This region occupies the volcanic plains to the North and West of Melbourne. Red wines particularly Shiraz are the stand out wines of this region, but there is potential for Italian red varieties and Tempranillo here.

- Arundel, Bacchus Hill, Galli Estate, Goona Warra Vineyard, Longview Creek, Wildwood, Witchmount Estate, Sweet Water Hill Wines

Yarra Valley

This region achieved worldwide renown as "Lilydale" during the nineteenth century. It is now regarded by some as Victoria's premium wine region. The climate varies with aspect and altitude but the region has a climate that is generally to the cooler end of the spectrum, at least by Australian standards. Red, white and sparkling wines are produced here and the proximity to Melbourne and scenic beauty has encouraged many wineries to expand their operations to include cafes, restaurants and luxury accommodation.

- Allinda, Badgers Brook Yarra Valley, Bianchet, Boat O'Craigo, Brammar Estate, Brumfield, Bulong Estate, Carlei Estate, Coldstream Hills, Coombe Farm Vineyard, Copper Bull, Diamond Valley Vineyards, Evelyn County Estate, Giant Steps, Helen's Hill Estate, Henley Hill, Hoddles Creek Estate, Immerse, Innocent Bystander, Jamsheed, Kellybrook, Killara Estate, Lillydale Estate, Mac Forbes Wines, Maddens Rise, mas serrat, Metier Wines, Millers Dixons Creek Estate, Oakridge, Outlook Hill, Punt Road, Redbox (Yarra Valley), Rochford Wines, Roundstone Winery, Samson Hill Estate, Seville Estate, Seville Hill, St Huberts, Steels Creek Estate, Stefani Estate, Sticks, Sutherland Estate, The Wanderer, Tokar Estate, Toolangi Vineyard, Whitsend Estate, Wild Cattle Creek Winery, Yarra Burn, Yarra Glen, Yarra Ridge, Yarra Yarra, Yarra Yering, Yarraloch, Yarrawood, Yering Station, Yeringberg

WESTERN VICTORIA ZONE

This Zone contains three formally recognized wine regions, plus the unofficial area of Ballarat. The climate varies from cool windy uplands around Ballarat to warmer dry areas in The Pyrenees and Grampians and (relatively) wet and cool maritime conditions in Henty.

Ballarat

The climate here is cool and the region shares some characteristics of the Nearby Macedon Ranges region. Pinot Noir and Chardonnay are grown for Sparkling and still wines. The original Yellowglen Winery is located here but the brand name is no longer associated with the region. Most of the wineries are quite small and few seem interested in alternative varieties.

- Whitehorse Wines

Grampians

The region was formerly known as Great Western, which is a small village on the Western Highway. Two wineries have dominated the area since the 1860s, Best's and Seppelt Great Western. The region now produces excellent red wines but for many years Great western was synonymous with sparkling wines - or champagne as we called it back then. Seppelt Sparkling Burgundy was also produced here, it is now called Sparkling Shiraz and the style has been emulated elsewhere.

A feature of the region is Best's nursery block. It was planted with a fruit salad of varieties in the nineteenth century and still yields a small crop from 150 year old vines. A few of the varieties in the block have not been identified; they are possibly the only vines of obscure varieties left in the world, the rest being lost to *Phylloxera*.

- Bests, Cathcart Ridge Estate, Donovan Wines, Jillian Wines, Michael Unwin Wines, Mount Cole Wineworks, Mount Langi Ghiran Vineyards, Seppelt Great Western, The Gap, Varrenti Wines, Westgate Vineyard

Henty

This little known region is located in Victoria's south west corner, just across the Border from the more famous Coonawarra. The climate is cool and is especially suitable for early ripening varieties Pinot noir, Meunier Chardonnay and Riesling. Seppelt Drumborg and Crawford River wineries were established in the 1970s, surprisingly few other wineries have followed. The region has obvious potential for Pinot gris which is grown in the area but mostly vinified elsewhere.

- Bochara, Mount Pierrepoint Estate, Rowans Lane Wines

Pyrenees

It takes a good imagination to see any resemblance between this region and the mountains of the same name which from Spain and France. The modern era of viticulture in the region dates from 1963 when Chateau Remy was established to produce brandy. It was soon realised that the region had potential for red wine production and a steady stream of new vineyards and wineries have been established since the 1970s. Shiraz is the most commonly grown variety for red wines, Pinot Noir and Chardonnay are grown for sparkling wines. Among the alternative varieties Tempranillo seems to have the most potential. A few wineries are also using Italian varieties.

- Counterpoint Vineyard, Dogrock Winery, Eleven Paddocks, Macketh House Historic Vineyard, Massoni, Melross Estate, Mount Avoca, Peerick Vineyard, Polleters Vineyard, Pyren Vineyard, Quoin Hill, Romantic Vineyard, Shays Flat Vineyard, St Ignatius Vineyard, Summerfield, Taltarni, Warrenmang Vineyard

NEW SOUTH WALES

Two regions dominate the wine scene in NSW, the Hunter Valley for its role as the birthplace of Australian wine and the Riverina which produces huge volumes. But over recent years some newer regions have stated to assert themselves.

BIG RIVERS ZONE

There are two regions in this Zone the Riverina and Perricoota, both irrigation districts. The sole winery of interest in the Zone outside the regions is at Walla Walla, north of Albury. It has more in common with Rutherglen than the Riverina

- Walla Wines

Perricoota

This is a tiny region around the township of Moama on the Murray River. The hot dry summers make irrigation essential. The region is very new and seems destined for the production of classic varieties for inclusion in large volume wines. There is potential for boutique wines to be produced for the Echuca tourist market just over the river.

- Moama Wines, Morrisons Riverview Winery, Redbox Perricoota, Silverfox Wines, St Annes Vineyards, Stevens Brook Estate

Riverina

This region produces about three times as much wine as the rest of the state put together, much of it is destined for entry level wines for export. The climate is hot and all vines are irrigated. Red, white and fortified styles are made, usually with classic varieties but there are some innovators in the region. There is more Durif grown in the region than most people realise and Westend Estate makes a quality wine from the obscure variety Saint Macaire.

- Beelgara Estate, Berton Vineyards, Casella, Clancy's of Conargo, Dayleswood Winery, De Bortoli, Lillypilly Estate, McWilliams, Melange Wines, Miranda Wines, Griffith, Piromit Wines, Red Mud, Riverina Estate Wines, Terrel Estate Wines, Vico, Wagga Wagga Winery, Warburn Estate, Westend Estate, Yarran

CENTRAL RANGES ZONE

The Central Ranges Zone is on the inland side of the Great Dividing Range. The climate is cool in the higher more rugged eastern part of the Zone, on the western side the topography becomes increasingly flat and the climate hotter and drier. Some wineries around Bathurst may eventually become incorporated into a formally recognised region.

- Barton Creek, Bell River Estate, Chateau Champsaur, Monument Vineyard, Sand Hills Vineyard, Stockman's Ridge, Vale Creek Wines, Winooka Park

Cowra

This region occupies broad valleys on the inland side of the Great Dividing Range. The climate is moderate to warm. Much of the wine produced here is blended by big companies into their mass produced brands, but there are some smaller producers. Shiraz and Chardonnay are the dominant varieties, but some interest is now being shown in expanding the range. The warmish climate indicates that Petit Verdot, Sangiovese and Tempranillo could play a bigger role in the future.

- Catherine's Ridge, Hamiltons Bluff, Kalari Wines, Mulligan Wongara Vineyard, Mulyan, Wallington Wines, Windowrie Estate

Mudgee

This is yet another wine region where the wine industry was established in the gold era (1850s) and faded away after a boom to be revived in the 1970s. There are now many large and small vineyards, the larger ones are part of multi regional companies who make wine, bottle or distribute it from facilities elsewhere.

The climate in Mudgee is cooler than the Hunter Valley just over the range but still warm by Australian standards. The wines are predominantly reds from Shiraz and Cabernet Sauvignon but many of the smaller wineries are using Italian and Spanish varieties.

- 5 Corners Wines, Andrew Harris Vineyards, Blue Wren, Botobolar, Burnbrae Winery, Clearview Estate Mudgee, Creeks Edge Wines, De Beaurepaire Wines, de Mestre Wines, Di Lusso Estate, Elliot Rocke Estate, Frog Rock, Gelland Estate, Jamieson Estate, Lawson Hill, Logan Wines, Louee Wines, Lowe Family Wines, Mansfield Wines, Miramar, Mountilford, Mudgee Growers, Mudgee Wines, Oatley Wines, Optimiste, Petersons Glenesk Estate, Pieter van Gent, Poet's Corner, Prince Hill Wines, Robert Stein, Seldom Seen, Shawwood Estate, Skimstone, Vinifera Wines, W Wine of Mudgee, Wells Parish Wines

Orange

This region has a variable climate, depending on altitude, but it does have the coolest overall environment. In fact the climate on the slopes of Mt Canobalas was identified by viticulturalist John Gladstones as one of the best areas for cool continental climate but with the topography for cold air drainage to avoid frost damage. Red wines made from Shiraz and Chardonnay based whites are common here with a few of the smaller vineyards dabbling in alternative varieties. It is surprising that Tempranillo seems to be absent from the region.

- Angullong Wines, Belgravia Vineyards, Bloodwood, Borrodell on the Mount, Canolobas-Smith, Cargo Road Wines, Cumulus Wines, Ibis Wines, Jarrets of Orange, Mayfield Vineyard, Orange Mountain, Orchard Road, Pepper Tree Wines, Philip Shaw, Prince of Orange, Printhie Wines, Ross Hill Wines, Sharpe Wines of Orange, Toogoolah Wines, Word of Mouth Wines

HUNTER ZONE

This zone includes the region known as Hunter valley which has the subregion Broke Fordwich and several others in the making. There is an argument, put forward by James Halliday among others, that there should be two regions, the Upper Hunter and Lower Hunter but the GI Committee seems to think otherwise.

The climate, especially in the lower parts of the region, is warm and humid with summer rainfall and on the face of it unsuitable for viticulture. However over 150 years of very successful winemaking has shown that vines are perhaps more hardy than we think. The two standout varieties in this region are Semillon and Shiraz, but there is considerable interest in other varieties. Verdelho and Chambourcin are capable of producing well in conditions where other varieties would succumb to disease. Barbera is also grown in the region.

- Adina Vineyard, Allandale, Allyn River Wines, Apthorpe Estate, Arrowfield, Audrey Wilkinson, Ballabourneen Wines, Belgenny Vineyard, Bell's Lane Wines, Benwarin Wines, Beyond Broke Vineyard, Bimbadeen Estate, Bimbadgen Estate, Bishop Grove Wines, Boatshed Vineyard, Briar Ridge Vineyard, Broke's Promise, Broke Estate, Brokenwood Wines, Brown's Farm Winery, Brush Box Vineyard, Calais Estate, Camp Road Estate, Camyr Allyn Wines, Capercaillie, Catherine Vale Vineyard, Chislehurst Estate, Colvin Wines, Constable Vineyards, Cooper wines, Cruickshank Callatoota Wines, David Hook Wines, De Iulius, Dimbulla, Draytons, Drews Creek Wines, Echo Ridge Wines, Elsmore's Caprera Grove, Elysium Vineyard, Emmas Cottage Vineyard, Ernest Hill Wines, Fairview Wines, Farrells Limestone Creek, First Creek, Foate's Ridge, Fordwich Estate, Gabriel's Paddocks Vineyard, Gartelmann Hunter, Ghost Riders Vineyard, Glandore Estate, Glendonbrook, Glenguin, Golden Grape Estate, Heartland Vineyard, Hillside Estate, Hollyclare, Honeytree Estate, Hope Estate, House of Certain Views, Hudson's Peake Wines, Hungerford Hill, Idle Hands Wines, Idlewild, Iron Gate Estate, Ironbark Hill Estate, Ivanhoe Wines, James Estate, Juul Wines, Keith Tulloch Wine, Kevin Sobels Wines, Krinklewood, Kulkunbulla, Little's Winery, Little Wine Company, Lucy's Run, Mabrook Estate, Madigan Vineyard, Margan Family, McLeish Estate, Meera Park, Milldale Estate Vineyard, Molly's Cradle, Molly Morgan Vineyard, Monahan Estate, Moorebank Vineyard, Mount Broke Wines, Mount Eyre Vineyards, Mount View Estate, Mount Vincent Estate, Nightingale Wines, Oakvale, Outram Estate, Palmers Wines, Pangallo Estate, Penmara, Piggs Peake Winery, Pokolbin Estate, Polin & Polin, Pothana, Pyramid Hill Wines, Racecourse Lane Wines, Reg Drayton, Ridgeview Wines, Robyn Drayton, Rosebrook Estate, Rothbury Ridge, Saddlers Creek Wines, Sandalyn Wilderness Estate, Serenella Estate, Sevenoaks Wines, SmithLeigh Vineyard, Southern Grand Estate, Splitrock Vineyard Estate, St Petrox, Stirling Wines, Stomp, Stonehurst Cedar Creek, Tallavera

Grove Winery, Tamburlaine, Telgherry, Tempus Two, Tinonnee Vineyard, Tintilla Wines, Tower Estate, Tulloch, Two Rivers, Tyrrells, Undercliff, Vercoes Vineyard, Verona Vineyard, Vinden Estate, Wandin Valley Estate, Warraroong Estate, Wattlebrook Vineyard, Wild Broke Wines, Winbourne Wines, Windsors Edge, Wright Family Wines, Yarraman Estate

NORTHERN RIVERS ZONE

This Zone stretches along the NSW Coast from Newcastle up to the Queensland border. It includes the Hastings River Region. The warm and humid weather with plenty of summer and vintage time rain provides a challenge. Many wineries use the French hybrid variety Chambourcin for its resistance to fungal diseases. Verdelho, while not a hybrid is also a feasible variety for white wine in these areas.

- Divers Luck Wines, Friday Creek Resort, Great Lakes Wines, Inlam Estate, Red Tail, Two Tails Wines, Villa d'Esta Vineyard, Alderley Creek Wines Estate, Raleigh Wines, Stroud Valley Wines, Wallambah Vale Wines, Wonganella Wines

Hastings River

This region is centers around the coastal city of Port Macquarie. John Cassegrain has pioneered viticulture in this region by developing techniques and practices to suit the local environment and encouraging others to set up vineyards. The problem with the region is summer rainfall and humidity making disease control essential.

- Bago Vineyards, Cassegrain, Douglas Vale, Inneslake, Long Point Vineyard, Sherwood Estate

NORTHERN SLOPES ZONE

This zone in Northern NSW covers a large area of the inland slopes of the Great Dividing Range. In 2008 an area in the middle of the Zone, from Tenterfield in the north to Tamworth in the south, was declared as the New England Wine Region. A few wineries missed lie outside the new region and are still listed in the Zone.

- Reedy Creek, Tangaratta Estate

New England Wine Region

This region was declared in January 2008. The climate here is marked by cold winters and moderate to warm summers and dry autumn, ideal conditions for wine growing. The landscape offers many opportunities for site selection to grow quality vineyards and we can expect a steady increase in quality wines from the regions local experience and expertise grows. On the face of it seems ideally suited to Tempranillo, but many other varieties will find a home here.

- Doctors Nose Wines, Dumaresq Valley Vineyard, Kitty Crawford Estate, Kurrajong Downs, Laurellyn Wines, Melville Hill Estate Wines, Mihi Creek Vineyard, Richfield Estate, Toppers Mountain, Walden Woods Farm, Whyworry Wines, Wright Robinson of Glencoe

SOUTH COAST ZONE

This Zone includes all of New South Wales south of Newcastle and east of the Great Dividing Range. It contains two declared regions and a few wineries scattered throughout the rest of the area. The climate is influenced by the proximity to the Pacific Ocean for while it moderates the extremes of temperature it also brings humidity and unwelcome rain during summer and vintage.

- Belgrave Park Winery, Cobbitty Wines, Disaster Bay Wines, Grevillea Estate, Lyrebird Ridge Organic Winery, Rusty Fig Wines, Tilba Valley, Tizzana Winery

Shoalhaven Coast

This region includes a cluster of wineries around the many seaside tourist towns south of Kiama. The climate is not ideal but a steady stream of tourists through the cellar door means that the average to good wines can be sold. Verdelho and Chambourcin are grown as reliable producers in wetter years.

- Bawley Vale Estate, Cambewarra Estate, Coolangatta Estate, Crooked River Wines, Cupitt's Winery, Fern Gully Winery, Humphries Estate, Jasper Valley, Kladis Estate, Mopoke Ridge Winery, Nowra Hill Vineyard, Roselea Estate, Seven Mile Vineyard, The Silos Estate, Yarrawa Estate

Southern Highlands

At an altitude of 600-750 metres the vineyards in this region enjoy a cool climate. The region is still developing, for wine as well as a venue for country retreats and tourism from nearby Sydney. This is one area of New South Wales that has embraced Pinot gris. Among the newer red varieties Tempranillo and Barbera have produced the best results to date.

- Artemis Wines, Berrima Estate, Blue Metal Vineyard, Bousaada, Centennial Vineyards, Cushendell, Cuttaway Hill Estate, Eling Forest Winery, Farago Hill, GrumbleBone Estate Vineyard, Joadja Vineyards, McVitty Grove, Mount Ashby Estate, Mundrakoona Estate, Ruane Winery, Saint Derycke's Wood, Southern Highland Wines, Tertini Wines

SOUTHERN NEW SOUTH WALES ZONE

There are four regions in this Zone, and a few wineries which fall outside the declared regions. The Zone encompasses the NSW part of the Australian Alps as well as adjacent foothills and valleys on the inland side.

- Kingsdale Wines, Snowy Vineyard, Transylvania Winery

Canberra District

Although the city is in the ACT most of the wineries are in the state of NSW. The climate is quite cool and continental. Among the classic varieties Riesling is a good variety to look for and Clonakilla's Shiraz Viognier is now reaching cult status. There are a few wineries stepping out of the square with new varieties, for example Lerida Estate with Australia's first Gruner Veltliner, and Mount Majura have established that Tempranillo goes very well in these parts.

- Barton Estate, Brindabella Hills, Capital Wines, Clonakilla, Doonkuna Wines, Even Keel Wines, Four Winds Vineyard, Jeir Creek, Kamberra, Lambert Vineyards, Lark Hill Winery, Lerida Estate, Little Bridge, Long Rail Gully, Madew Wines, McKellar Ridge, Mount Majura, Mundoonen, Murrumbateman Winery, Pialligo Estate, Quarry Hill Wines, Ravensworth Wines, Shepherds Run, Surveyor's Hill Winery, Taemus Wines, Yarrh Wines, Yass Valley Wines

Gundagai

This region straddles the Hume Highway around the famous dog-on-the-tuckerbox town. The topography varies from mountainous in the south east to gentler slopes to the north and west. The wineries in the region are new and the region is yet to establish a reputation.

- Bidgeebong

Hilltops

This region, centred on the township of Young produces mainly red wines but there are also some good Rieslings produced here as well. A couple of wineries are working outside the square with some exotic varieties including Corvina and Rondinella.

- Ballinaclash Wines, Barwang, Binbilla, Freeman Vineyards, Grove Estate Wines, Moppity Vineyards, Trandari, Woodonga Hill

Tumbarumba

This is a remote wine region in the Australian Alps. There are only a few wineries in this very cold region, a couple of them are using alternative varieties.

- Lankeys Creek Wines, Lighthouse Peak

WESTERN PLAINS ZONE

This vast Zone is really the 'what's left' part of New South Wales. There is a cluster of vineyards around Dubbo on the fringes of the Central Ranges Zone, which may eventually become the Macquarie Valley Region. Ken Borchardt at Red Earth Estate is trialling a range of varieties to find some suitable for the local environment. The rest of the area is too hot and dry for viticulture.

- Boora Estate, Canonbah Bridge, Lazy River Estate, Macquarie Grove Vineyards, Red Earth Estate, Tombstone Estate

WESTERN AUSTRALIA

The wine regions are in this state are divided, rather farcically, into five Zones. One Zone, the Eastern Plains, Inland and North of Western Australia Zone covers about a third of the continent. It contains just two wineries. Another, the West Australian South East Coastal Zone has just one. The action, and indeed the population of WA, is in the other three Zones.

CENTRAL WESTERN AUSTRALIAN ZONE

This Zone occupies a 600km long Strip of land generally 50-150km inland from the Indian Ocean Coast. The land is mainly used for wheat production and grazing but a few small vineyards are making the most of the strongly continental climate.

- Chapman Valley Wines, Chidlows Well, Halina Brook, Jylland Vineyard, Lamonts, Wandoo Farm

GREATER PERTH ZONE

This Zone extends north and South of Perth along the coastal plain and it extends over the Darling Ranges. It has three regions, Peel, Perth Hills, and Swan District with the Swan Valley making up a subregion of the Swan District.

Peel

The coastal section of this region is around the rapidly growing satellite city of Mandurah. It has a classic Mediterranean climate. Inland the altitude of up to 300 metres provides some relief from the summer heat. Surprising for such a warm region the majority of the wine produced here is white, based on Chenin blanc, Chardonnay and Verdelho. Shiraz is the dominant red variety.

- Amarillo Wines, Cape Bouvard, Concotton Creek, Drakesbrook Wines, Emmetts Crossing Wines, Hotham Ridge Winery, Peel Estate, Peel Ridge, Stakehill Estate, The Ritual, Tuart Ridge, Wandering Brook Estate, Wandering Lane

Perth Hills

The topography of the region is a maze of small hills and valleys with a variety of microclimates. Altitudes, and therefore summer temperatures vary but some sites have proved suitable for Pinot noir and sparkling wines. Most wineries are small and all are relatively new so experimentation with varieties and styles is continuing.

- Briery Estate, Carosa, Chittering Valley Winery, Cosham, Darlington Estate, Francois Jacquard, Hainault, Hartley Estate, Jarrah Ridge Winery, Kyotmunga Estate, Lake Charlotte Wines, Lion Mill Winery, McCuskers Vineyard, Millbrook Winery, Myattsfield

Vineyard and Winery, Parkerville Ponds Vineyard, Stringybark, Walsh Family Winery, Western Range Wines, Windshaker Ridge

Swan District

This region extends north of Perth mainly along the coastal plain. The temperature is generally hot with some relief given by breezes from the Indian Ocean. The Swan Valley sub-region occupies the South east corner of region. White wines, based on the old WA favourites Chenin blanc and Verdelho are made here as well as shiraz based reds.

- Bella Ridge Estate, Berrigan Wines, Carabooda Estate, Cheriton, Faranda Wines, Franand Wines, Olive Farm, Paul Conti Wines, Riseborough Estate, Sussanah Brook Wines, Valley Wines

Swan Valley

This is one of Australia oldest wine regions, dating back over 170 years. The region is very hot but makes more white wine than red, with Chardonnay joining Chenin blanc and Verdelho, often in three way blends. Reds are generally based on Shiraz and there are some impressive fortified wines made as well. Many of the larger wineries use material from cooler WA regions for winemaking here.

- Ambrook Wines, Black Swan Winery, Carilley Estate, Charlies Estate Wines, Faber Vineyard, Fish Tail Wines, Garbin Estate, Harris Organic Wines, Heafod Glen Winery, Houghton, Jane Brook Estate, John Kosovich Wines, Lancaster Wines, Lilac Hill Estate, Little River Wines, Mann, Moondah Brook, Neilson Estate Wines, Oakover Estate, Pinelli, Riverbank Estate, Sandalford Wines, Sittella, Swan Valley Wines, Swanbrook Estate Wines, Talijancich, The Natural Wine Company, Upper Reach Vineyard, Vino Italia, Westfield, White's Vineyard, Windy Creek Estate

SOUTH WEST AUSTRALIA ZONE

This Zone is where most of the action is in WA. It contains six regions, one of which, Great Southern, has five sub-regions. The influence of the Southern and Indian Ocean on temperature and rainfall is felt in all of the regions but some are quite cool.

Blackwood Valley

This region lies inland from Margaret river and has a somewhat more continental climate. Much of the grape harvest in the region, mostly from red wine varieties, is sold for vinification elsewhere.

- Blackwood Wines, Tanglewood Vines, Wattle Ridge Wines

Geographe

This region is centred on the coastal City of Bunbury and it draws its name from Geographe Bay. Vineyards extend from the coastal plains up into the valleys of the Darling Range. Capel Vale is the best known and biggest winery in the region but there are many more smaller wineries pushing the boundaries with a range of varieties.

- Barrecas, Blackboy Ridge Estate, Brookhampton Estate, Byramgou Park, Capel Vale, Carlaminda Estate, Donnybrook Estate, Ferguson Falls Winery, Geographe Wines, Hackersley, Harris River Estate, Idylwild Wines, Kotai Estate, Mandalay Road, Mazza, Myalup Wines, Pepperilly Estate Wines, Smallwater Estate, Thomson Brook Wines, Vineyard 28, Whicher Ridge, Willow Bridge Estate, Wordsworth Wines, Wovenfield, Yokain Vineyard Estate

Great Southern Region

This region is centred on the Port City of Albany and faces the Southern Ocean. It comprises five Subregions, Albany, Denmark, Frankland River, Mount Barker and Porongurup. The climate varies with the distance from the coast but generally this is the coolest region in Western Australia. Much of the land area and a few of the vineyards are outside of the subregions.

- Jingalla, Middlesex 31, The Lily Stirling Range

Albany Subregion

This region is situated around the historic whaling station and naval and military base of Albany. The climate is cool enough for Pinot Noir which along with other classic varieties makes up the majority of the crush. There seems to be little interest in growing alternative varieties, even Riesling is rare here.

- Kalgan River Wines

Denmark Subregion

The town of Denmark, and the wine region is further to the west from Albany. The region is cool and wet by WA standards. The wineries here seem wedded to classic varieties, the exception being West Cape Howe who use Viognier very well.

- West Cape Howe Wines

Frankland River Subregion

This is the most inland of the Subregions in Great Southern. It is fairly remote and many of the vineyards supply grapes for use by companies based outside the region. The Riesling produced here is among the best in Australia but again there seems to be little interest in alternative varieties.

- Alkoomi, Ferngrove Vineyards, Frankland Estate

Mount Barker Subregion

This region is located far enough inland to have a climate approaching continental. It is also cool enough for varieties such as Riesling, Gewurztraminer and Muller Thurgau, along with the ubiquitous Shiraz, Cabernet and Chardonnay.

- Chatsfield, Fox River Wines, Galafrey, Goundrey, Plantagenet, Poachers Ridge Vineyards, Trappers Gully

Porongurup Subregion

Immediately to the east of Mt Barker this region has a similar climate but is distinctive topography sets it apart. Excellent Riesling is made here and a few wineries are using alternative varieties.

- Duke's Vineyard, Fernbrook Estate, Montefalco Vineyard, Mount Trio Vineyard

Manjimup

This region has a similar climate to Margaret River with a more pronounced continentality. The varietal profile is fairly mainstream but a few wineries have taken some tentative steps to varietal diversity with Sangiovese and Viognier.

- Chestnut Grove, Peos Estate, Piano Gully, Stone Bridge Estate, Yanmah Ridge

Margaret River

This region is no doubt the jewel in the crown of Western Australian wine. The climate is described as 'Mediterranean' but the winters are very mild and summers don't reach the scorching heat experienced elsewhere. The high rainfall falls almost entirely during the five months between M may and September. The topography and soil type vary. A major factor in vineyard site selection is protection from the strong winds blowing from the neighbouring oceans. The varieties grown here reflect the idea that the region could be Australia's answer to Bordeaux, but of course there is Chardonnay and Shiraz here as well. As the list below demonstrates there are many wineries that have broken out of that mold, many with the WA staples of Verdelho and Chenin blanc but others have strayed even further from orthodoxy.

- Abbey Vale, Adinfern, Amberley Estate, Arimia Margaret River, Arlewood Estate, Ashbrook Estate, Barwick Wines, Beckett's Flat, Blue Manna, Blue Poles Vineyard, Brairose Estate, Briarose Estate, Brookwood Estate, Broomstick Estate, Cape Grace

Wines, Cape Mentelle, Celestial Bay, Chalice Bridge Estate, Chapman's Creek Vineyard, Churchview Estate, Cullen Wines, Deep Woods Estate, Eagle Vale, Evans and Tate, Fermoy Estate, Flinders Bay, Flying Fish Cove, Forester Estate, Gherardi Wines, Green Valley Vineyard, Happs, Harman's Ridge Estate, Hay Shed Hill Wines, Heron Lake Estate, Higher Plane, Howard Park Wines, Island Brook, Jarvis Estate, Jerusalem Hollow, Juniper Estate, Karri Grove Estate, Kneedeep, Knotting Hill Vineyard, MadFish Wines, Marri Wood Park, Marybrook Vineyards & Winery, McHenry Hohnen, Miles from Nowhere, Moaning Frog, Moss Brothers, Olsen, Pennan, Plan B, Redgate, Rivendell, Settlers Ridge, Seven Ochres Vineyard, Stella Bella, Stellar Ridge, Swings & Roundabouts, Swooping Magpie, Tassell Park Wines, The Grove Vineyard, Vasse River Wines, Victory Point Wines, Virage , Voyager Estate, Watershed Wines, Willespie, Wills Domain Vineyard, Windance Wines, Windows Margaret River, Wise Wine, Wombat Lodge, Woodlands, Woody Nook

Pemberton

This region is characterised more by its forests of giant Eucalypts than by its vineyards. The climate is cool and wet, combined with fertile soils this leads to vigorous growth. White wines from Chardonnay and the other WA favourites of Chenin Blanc and Verdelho are grown here as well as Cabernet and Merlot based reds.

- Angelicus, Bellarmine Wines, Black George, Channybearup, Donnelly River Wine, Fonty's Pool Vineyards, Lillian, Merum, Phillips Estate, Salitage, Smithbrook, Tantemaggie, Truffle Hill Wines

QUEENSLAND

There are only two formally recognised regions in Queensland, the Granite Belt and South Burnett; many southerners are surprised to know that there is any wine there at all. There are several other small clusters of vineyards scattered about that might eventually become regions.

The following wineries are not in other regions mentioned below

- Brierley Wines, Governor's Choice Winery, Jimbour Wines, Kooroomba, Normanby Wines, Paradine Estate, Riversands Winery, Romavilla, Sarabah Estate, Stanton Estate, Three Moon Creek, Toowoomba Hills Estate, Waratah Vineyard, Warrego, Winya Wines

Darling Downs

There are more than a dozen wineries here but its overall production means that Darling Downs is still not formally recognised. The vineyards are situated in the hills and valleys around the city of Toowoomba. Conditions are not vastly different to those in the nearby Granite Belt and it is feasible that considerable expansion of viticulture could occur here. Rimfire Vineyards have been trialling a range of varieties to see how they respond to the local environment.

- Gowrie Mountain Estate, Lilyvale Wines, Rangemore Estate, Rimfire Vineyards, Wedgetail Ridge Estate

Granite Belt

This region abuts the New South Wales Border and the New England wine region. The altitude of the vineyards - 800m plus - and its inland location gives the region a decidedly continental climate. The wines made in this region stand on their merits they don't rely on any novelty value or depend on the tourist trade. Red and white wines do well here, with considerable interest in Tempranillo and Nebbiolo. Many of the local wineries in the Granite Belt are very interested in alternative varieties. They have formed a group to publicise them via map to find 'Strange Bird' wines.

- Ambar Hill, Aventine Wines, Back Pocket, Bald Mountain, Ballandeen Estate, Banca Ridge, Bents Road, Boireann, Bungawarra, Casley Mount Hutton Winery, Catspaw Farm, Cypress Post, Felsberg Winery, Golden Grove Estate, Granite Ridge Wines, Harrington Glen Estate, Heritage Estate, Hidden Creek, Jester Hill Wines, Kominos Wines, Lucas Estate, Mary Byrnes Wines, Mason Wines, Old Caves, Preston Peak, Pyramids Road Wines, Ravens Croft Wines, Ridgemill Estate, Robert Channon, Rumbarella, Severn Brae Estate, Stone Ridge, Summit Estate, Symphony Hill Wines, Tobin Wines, Whiskey Gully Wines, Windemere Wines, Winewood, Witches Falls Winery

South Burnett

This region is located north west of Brisbane around the regional centre of Kingaroy. The climate is quite hot but without extremes in summer. Verdelho and Chambourcin, two varieties of choice in hot humid areas, are prominent among the alternative varieties.

- Barambah Ridge Winery, Bridgeman Downs Cellars, Clovely Estate, Crane Wines, Dusty Hill Estate, Hunting Lodge Estate, Kingsley Grove, Moffatdale Ridge, Mount Appallan Vineyards, Rodericks, Stuart Range Estate, Tipperary Estate, Whistle Stop Wines

Queensland Coastal

There are a number of wineries along the coast, or in the immediate hinterland north and south of Brisbane. They face great difficulty in growing classic wine grape varieties. They get around this by using suitable varieties and in some cases by bringing in grapes or wine from elsewhere. Many of these wineries cater to the booming tourist market with restaurants and tasting rooms to give visitors a vineyard experience rather than being serious winemaking ventures.

- Albert River, Canungra Valley, Cedar Creek Estate, Eumundi Winery, Gecko Valley, Gin Gin Wines, Maleny Mountain Wines, Maroochy Springs, Mount Tamborine, Noosa Valley Winery, Norse Wines, O'Regan Creek Vineyard and Winery, Oceanview Estates, Settlers Rise Montville, Sirromet, Springbrook Mountain Vineyard, Tamborine Estate Wines, Twin Oaks, Wonbah Estate, Woongoroo Estate

TASMANIA

There are no formal regions in Tasmania, the whole state is a Zone. It is feasible to consider two or three regions, Northern, Southern and perhaps Eastern Tasmania. James Halliday divides the state into northern and southern regions and I follow suit.

A few large companies have vineyards in Tasmania for cool climate wines and material for sparkling wines, but for the most part Tasmania is the realm of boutique wineries.

Northern Tasmania

The greatest concentration of vineyards is in the Tamar Valley and Pipers River area. The varieties grown are generally Pinot noir and a range of different whites, the most interesting of which are Riesling, Pinot gris and Gewurztraminer. Pinot Noir and Chardonnay are used for the production of sparkling wines the best of which vie with those on the mainland. Special site selection is required for successful cultivation of other red varieties.

- Barringwood Park, Bass Fine Wines, Bay of Fires, Chartley Estate, Frogmore Creek Vineyard, Ghost Rock, Grey Sands, Hillwood Vineyard, Iron Pot Bay Wines, Jinglers Creek, Josef Chromy Wines, Marions Vineyard, Pipers Brook Vineyard, Pirie Estate, Rosevears Estate, St Matthias, Tamar Ridge, Three Willows Vineyard, Velo Wines, White Rock Vineyard, Wilmot Hills Vineyard

Southern Tasmania

The climate in this part of the state is of course quite cold and Pinot noir is the only red variety planted to any extent, but there are plenty of white varieties used. There are several clusters of vineyards but none produce a sufficient volume to be regarded as a formal region.

- 572 Richmond Road, Bracken Hill, Bream Creek Vineyard, Cape Bernier Vineyard, Craigow, Cross Rivulet, Elsewhere Vineyard, Freycinet, Grandview Vineyard, Herons Rise, Home Hill, Kraanwood, Meadowbank Estate, Milton Vineyard, Moorilla Estate, No Regrets, Palmara, Pooley Wines, Roslyn Estate, Spring Vale Wines, Stefano Lubiano, Sugarloaf Ridge, Wellington, Yaxley Estate

Wineries

The wineries in this list are using one or more alternative varieties. Please be aware that wineries appear and disappear regularly and often change names and ownership. The grape varieties that they use also change, this is the pioneering end of the industry – nobody is standing still.

The list is based on a database that is used to maintain the Vinodiversity.com website. The data was current at April 2010.

Additions and corrections should be forwarded to dah@vinodiversity.com

A

1847 (Barossa Valley)
www.eighteenfortyseven.com
Petit verdot

5 Corners Wines (Mudgee)
www.5corners.biz
Gewurztraminer

572 Richmond Road (Southern Tasmania)
Gewurztraminer

919 Wines (Riverland)
www.919wines.com.au
Durif, Graciano, Muscadelle, Palomino, Petit manseng, Sangiovese, Savagnin, Tempranillo, Tinto Cao, Touriga, Vermentino

Ada River (Gippsland)
Pinot gris

Adelina Wines (Clare Valley)
www.adelina.com.au
Grenache

Adina Vineyard (Hunter Valley)
www.adinavineyard.com.au
Pinot gris

Adinfern (Margaret River)
www.adinfern.com
Malbec

Akrasi Wines (Central Victoria Zone)
www.akrasiwine.com.au
Marsanne

Alan and Veitch (Adelaide Hills)
robertjohnsonvineyards.com.au
Viognier

Albert River (Queensland Coastal)
www.albertriverwines.com.au
Viognier

Alderley Creek Wines Estate (Northern Rivers Zone)
www.alderleycreekwines.com.au
Chambourcin, Verdelho

Aldinga Bay (McLaren Vale)
www.aldingabay.com.au
Sangiovese, Barbera, Petit verdot, Nebbiolo, Ruby Cabernet, Verdelho, Vermentino, Viognier

Alkoomi (Frankland River)
www.alkoomiwines.com.au
Malbec, Petit verdot, Viognier

All Saints Estate (Rutherglen)
www.allsaintswine.com.au
Chenin blanc, Durif, Marsanne, Muscadelle, Orange muscat, Roussanne, Ruby Cabernet

Allandale (Hunter Valley)
www.allandalewinery.com.au
Chambourcin, Verdelho

WINERIES • A

Allies Wines (Mornington Peninsula)
www.allies.com.au
Viognier

Allinda (Yarra Valley)
www.allindawinery.com.au
Savagnin

Allusion Wines (Southern Fleurieu)
www.allusionwines.com.au
Viognier

Allyn River Wines (Hunter Valley)
Chambourcin

Alta Wines (Adelaide Hills)
www.altavineyards.com.au
Pinot gris

Amadio (Adelaide Hills)
www.amadiowines.com.au
Dolcetto, Grenache, Pinot gris, Sangiovese, Tempranillo

Amarillo Wines (Peel)
Grenache

Ambar Hill (Granite Belt)
www.ambarhill.com.au
Verdelho

Amberley Estate (Margaret River)
www1.amberleyestate.com.au
Chenin blanc

Ambrook Wines (Swan Valley)
www.ambrookwines.com.au
Chenin blanc, Grenache, Verdelho

Amicus (McLaren Vale)
www.amicuswines.com.au
Malbec

Amietta Vineyard (Geelong)
www.amietta.com.au
Carmenere, Lagrein

Amulet Vineyard (Beechworth)
www.amuletvineyard.com.au
Barbera, Nebbiolo, Orange muscat, Pinot blanc, Pinot gris, Sangiovese, Trebbiano, Viognier

Anderson Winery (Rutherglen)
www.andersonwinery.com.au
Chenin blanc, Durif, Muscadelle, Petit verdot, Pinot gris, Sangiovese, Tempranillo, Viognier

Andrew Harris Vineyards (Mudgee)
www.andrewharris.com.au
Verdelho

Andrew Peace Wines (Swan Hill)
www.apwines.com
Crouchen, Grenache, Malbec, Moscato, Mourvedre, Sagrantino, Sangiovese, Viognier

Angelicus (Pemberton)
www.angelicus.com.au
Tempranillo

Angoves Winery (Riverland)
www.angoves.com.au
Biancone, Colombard, Doradillo, Gewurztraminer, Lexia, Palomino, Petit verdot, Pinot gris, Rubired, Ruby Cabernet, Sylvaner, Tempranillo, Verdelho, Viognier

Angullong Wines (Orange)
www.angullong.com.au
Barbera, Marsanne, Pinot gris, Sangiovese, Verdelho, Viognier

Annapurna (Alpine Valleys)
www.annapurnaestate.com.au
Pinot gris

Annie's Lane (Clare Valley)
www.annieslane.com.au
Mourvedre

Apthorpe Estate (Hunter Valley)
www.apthorpe.com.au
Chambourcin

Arakoon (McLaren Vale)
www.arakoonwines.com.au
Grenache, Mourvedre, Viognier

Arimia Margaret River (Margaret River)
www.arimia.com.au
Grenache, Mourvedre, Petit verdot, Verdelho, Zinfandel

Arlewood Estate (Margaret River)
www.arlewood.com.au
Marsanne, Roussanne

Armstead Estate (Heathcote)
www.armsteadestate.com.au
Marsanne

Arrivo (Adelaide Hills)
arrivo.com.au
Nebbiolo

Arrowfield (Hunter Valley)
www.arrowfieldwines.com.au
Petit verdot, Verdelho

Artemis Wines (Southern Highlands)
www.artemiswines.com.au
Tempranillo

Artwine (Clare Valley)
www.artwine.com.au
Graciano, Grenache, Pinot gris, Savagnin, Tempranillo, Viognier

Arundel (Sunbury)
www.arundel.com.au
Viognier

Ashbrook Estate (Margaret River)
www.ashbrookwines.com.au
Verdelho

Ashton Hills (Adelaide Hills)
Gewurztraminer, Malbec, Petit verdot, Pinot gris

Audrey Wilkinson (Hunter Valley)
www.audreywilkinson.com.au
Gewurztraminer, Malbec, Tempranillo, Verdelho, Zinfandel

Austin's Wines (Geelong)
www.austinswines.com.au
Pinot gris, Viognier

Australian Domaine Wines (Clare Valley)
www.ausdomwines.com.au
Grenache

Australian Old Vine Wines (Riverland)
www.australianoldvine.com.au
Chambourcin, Colombard

Avalon Wines (King Valley)
www.avalonwines.com.au
Tempranillo, Sangiovese, Verdelho

Aventine Wines (Granite Belt)
www.aventinewines.com.au
Nebbiolo, Sangiovese

Avonmore Estate (Bendigo)
www.avonmoreestatewine.com
Sangiovese, Viognier

B

B3 Wines (Barossa Valley)
www.b3wines.com.au
Grenache, Mourvedre

Baarrooka (Strathbogie Ranges)
www.baarrooka.com.au
Petit verdot, Zinfandel

Bacchus Hill (Sunbury)
www.bacchushill.com.au
Fragola, Chenin blanc, Nebbiolo

Back Pocket (Granite Belt)
www.backpocket.com.au
Graciano, Tempranillo

Baddaginnie Run (Strathbogie Ranges)
www.baddaginnierun.net.au
Verdelho

Badgers Brook Yarra Valley (Yarra Valley)
www.badgersbrook.com.au
Viognier

Bago Vineyards (Hastings River)
www.bagovineyards.com.au
Chambourcin, Moscato paradiso, Petit verdot, Savagnin, Tannat, Verdelho, Viognier

WINERIES • B

Baie Wines (Geelong)
www.baiewines.com.au
Pinot gris

Bailey's of Glenrowan (Glenrowan)
www.baileysofglenrowan.com.au
Muscadelle

Baillieu Vineyard (Mornington Peninsula)
www.baillieuvineyard.com.au
Pinot gris, Meunier

Bald Mountain (Granite Belt)
www.baldmountainwines.com.au
Verdelho

Ballabourneen Wines (Hunter Valley)
www.ballabourneenwines.com.au
Verdelho

Ballandeen Estate (Granite Belt)
www.ballandeanestate.com
Chenin blanc, Malbec, Nebbiolo, Sylvaner, Viognier

Ballast Stone Estate (Currency Creek)
www.ballaststonewines.com
Grenache, Moscato, Petit verdot, Pinot gris

Ballinaclash Wines (Hilltops)
www.ballinaclash.com.au
Viognier

Balthazar (Barossa Valley)
www.balthazarbarossa.com
Grenache, Mourvedre, Viognier

Banca Ridge (Granite Belt)
www.usq.edu.au/qcwt/bancaridge
Marsanne

Banderra Estate (Central Ranges Zone)
Colombard, Mourvedre

Banks Road (Geelong)
www.banksroad.com.au
Pinot gris

Banrock Station (Riverland)
www.banrockstation.com.au
Fiano, Montepulciano, Moscato, Petit verdot, Savagnin, Tempranillo

Barambah Ridge Winery (South Burnett)
www.barambahridge.com.au
Verdelho

Barmah Park Wines (Mornington Peninsula)
www.barmahparkwines.com.au
Pinot gris

Barossa Valley Estate (Barossa Valley)
www.bve.com.au
Grenache

Barrecas (Geographe)
www.barrecas.com.au
Barbera, Sangiovese, Zinfandel

Barringwood Park (Northern Tasmania)
www.barringwoodpark.com.au
Meunier, Pinot gris, Schonburger

Barristers Block (Adelaide Hills)
www.barristersblock.com.au
Grenache, Mourvedre

Barrymore Estate (Mornington Peninsula)
www.barrymore.com.au
Pinot gris

Bartagunyah Estate (Southern Flinders Region)
www.smartaqua.com.au/bartagunyah
Viognier

Barton Creek (Central Ranges Zone)
www.nyranghomestead.com.au
Sangiovese

Barton Estate (Canberra)
www.bartonestate.com.au
Malbec, Petit verdot, Pinot gris, Sangiovese, Viognier

Barwang (Hilltops)
www.mcwilliams.com.au
Pinot gris

Barwick Wines (Margaret River)
www.barwickwines.com
Viognier

Barwon Ridge Wines (Geelong)
www.barwonridge.com.au
Marsanne

Basalt Ridge (Macedon Ranges)
Pinot gris

Basedow (Barossa Valley)
www.basedow.com.au
Grenache

Bass Fine Wines (Northern Tasmania)
Pinot gris

Bass Phillip (Gippsland)
www.bassphillip.com.au
Gamay

Bass River (Gippsland)
www.bassriverwinery.com
Pinot gris

Battely Wines (Beechworth)
www.battelywines.com.au
Counoise, Durif, Marsanne, Viognier

Battle of Bosworth Wines (McLaren Vale)
www.edgehill-vineyards.com.au
Viognier

Battunga Vineyards (Adelaide Hills)
Pinot gris, Viognier

Bawley Vale Estate (Shoalhaven Coast)
www.bawleyvaleestate.com.au
Arneis, Chambourcin, Verdelho

Bay of Fires (Northern Tasmania)
www.bayoffires.com.au
Gewurztraminer, Pinot gris

Bayview Estate Winery (Mornington Peninsula)
www.bayviewestate.com.au
Pinot gris

Beach Road (Langhorne Creek)
www.beachroadwines.com.au
Greco bianco, Fiano, Petit verdot

Beckett's Flat (Margaret River)
www.beckettsflat.com.au
Malbec, Verdelho

Beechtree Wines (McLaren Vale)
www.beechtreewines.com.au
Arneis, Marsanne, Petit verdot

Beechwood Wines (Goulburn Valley)
Verdelho

Beelgara Estate (Riverina)
www.beelgara.com.au
Aranel, Chambourcin, Durif, Gewurztraminer, Marsanne, Pinot gris, Petit verdot, Verdelho, Viognier

Belalie Bend (Southern Flinders Region)
www.belelie bend.com.au
Mourvedre, Sangiovese

Belgenny Vineyard (Hunter Valley)
Verdelho

Belgrave Park Winery (South Coast Zone)
www.belgravepark.com
Marsanne, Roussanne, Sangiovese, Viognier

Belgravia Vineyards (Orange)
www.belgravia.com.au
Gewurztraminer, Roussanne, Viognier

Bell's Lane Wines (Hunter Valley)
Verdelho

Bell River Estate (Central Ranges Zone)
www.bellriverestate.com.au
Grenache

Bella Ridge Estate (Swan District)
www.bellaridge.com.au
Chenin blanc, Grenache, Kyoho, Mourvedre, Tempranillo, Trebbiano

Bellarine Estate (Geelong)
www.bellarineestate.com.au
Durif, Pinot gris, Viognier

Bellarmine Wines (Pemberton)
www.bellarmine.com.au
Petit verdot

Bellbrae Estate (Geelong)
www.bellbraeestate.com
Moscato, Roussanne

Ben Potts Wines (Langhorne Creek)
www.benpottswines.com.au
Malbec

Bendigo Wine Estate (Bendigo)
www.bendigowineestate.com.au
Petit verdot, Verdelho

Bent Creek Vineyards (McLaren Vale)
www.bentcreekvineyards.com.au
Grenache

Bents Road (Granite Belt)
www.bentsroadwinery.com.au
Marsanne, Verdelho

Benwarin Wines (Hunter Valley)
www.benwarin.com.au
Chambourcin, Sangiovese, Verdelho

Bergamin Wines (King Valley)
www.bergamin.com.au
Pinot gris

Berrigan Wines (Swan District)
www.berriganwines.com.au
Verdelho

Berrima Estate (Southern Highlands)
www.berrimaestate.com.au
Arneis

Berton Vineyards (Riverina)
www.bertonvineyards.com.au
Chenin blanc, Pinot gris, Trebbiano, Viognier

Bests (Grampians)
www.bestswines.com
Cabernet franc, Dolcetto, Meunier

Bethany (Barossa Valley)
www.bethany.com.au
Grenache

Beyond Broke Vineyard (Hunter Valley)
www.wine2go.com.au
Verdelho

Bianchet (Yarra Valley)
www.bianchet.com.au
Gewurztraminer, Marsanne, Verduzzo

Bidgeebong Wines (Gundagai)
www.bidgeebong.com
Tempranillo, Verdelho

Big Hill Vineyard (Bendigo)
www.bighillvineyard.com
Verdelho

Bimbadeen Estate (Hunter Valley)
www.bimbadeen.com.au
Verdelho

Bimbadgen Estate (Hunter Valley)
www.bimbadgen.com.au
Pinot gris, Sangiovese, Verdelho, Viognier

Binbilla (Hilltops)
www.binbillawines.com
Viognier

Birthday Villa Vineyard (Macedon Ranges)
Gewurztraminer

Biscay Wines (Barossa Valley)
Grenache, Viognier

Bishop Grove Wines (Hunter Valley)
www.bishopgrove.com.au
Verdelho

BK Wines (Adelaide Hills)
www.bkwines.com.au
Gewurztraminer, Pinot gris

Black George (Pemberton)
www.blackgeorge.com
Verdelho

Black Swan Winery (Swan Valley)
www.blackswanwines.com.au
Cabernet franc

Blackbilly (McLaren Vale)
www.blackbilly.com
Grenache, Mourvedre, Tempranillo

Blackboy Ridge Estate (Geographe)
www.blackboyridge.com.au
Chenin blanc

Blackets (Adelaide Hills)
www.blackets.com.au
Gewurztraminer

Blackford Stable Wines (Adelaide Hills)
www.blackfordstable.com.au
Pinot gris, Sangiovese

Blackwood Wines (Blackwood Valley)
www.blackwoodwines.com.au
Chenin blanc, Ruby Cabernet, Verdelho

Blamires Butterfly Crossing (Bendigo)
www.butterflycrossing.com.au
Viognier

Blanche Barkly (Bendigo)
www.bendigowine.org.au
Refosco

Bleasdale (Langhorne Creek)
www.bleasdale.com
Malbec, Verdelho

Bloodwood (Orange)
www.bloodwood.com.au
Cabernet franc, Malbec

Blown Away (McLaren Vale)
www.blownaway.net.au
Grenache, Kerner

Blue Manna (Margaret River)
www.bluemanna.com.au
Chenin blanc, Verdelho

Blue Metal Vineyard (Southern Highlands)
www.bluemetalvineyard.com
Pinot gris, Petit verdot, Saperavi, Sangiovese, Viognier

Blue Poles Vineyard (Margaret River)
www.bluepolesvineyard.com.au
Tempranillo, Viognier

Blue Wren (Mudgee)
www.bluewrenwines.com.au
Verdelho

Bluestone Lane (Mornington Peninsula)
www.bluestonelane.com
Meunier

Boat O'Craigo (Yarra Valley)
www.boatocraigo.com.au
Pinot gris, Roter veltliner, Shiraz Viognier

Boatshed Vineyard (Hunter Valley)
Chambourcin, Verdelho

Bochara (Henty)
www.bocharawine.com.au
Meunier, Petit manseng

Bockman (Adelaide Hills)
Pinot gris

Boggy Creek Vineyards (King Valley)
www.boggycreek.com.au
Barbera, Pinot gris, Sangiovese

Bogie Man Wines (Strathbogie Ranges)
www.bogiemanwines.com
Lagrein, Prosecco, Tempranillo

Boireann (Granite Belt)
www.boireannwinery.com.au
Barbera, Grenache, Mourvedre, Nebbiolo, Petit verdot, Tannat, Viognier

Bonneyview (Riverland)
Petit verdot

WINERIES • B

Boora Estate (Western Plains)
www.boora-estate.com
Tempranillo

Borrodell on the Mount (Orange)
www.borrodell.com.au
Gewurztraminer, Meunier

Botobolar (Mudgee)
www.botobolar.com
Crouchen, Marsanne

Bottin Wines (McLaren Vale)
www.vignabottin.com
Barbera, Sangiovese

Bousaada (Southern Highlands)
www.bousaada.com
Tempranillo

Bowe Lees (Adelaide Hills)
Nebbiolo, Tannat

Bowmans Run (Beechworth)
Gewurztraminer

Box Grove Vineyard (Nagambie Lakes)
www.boxgrovevineyard.com.au
Prosecco, Roussanne, Vermentino, Viognier

Box Stallion (Mornington Peninsula)
www.boxstallion.com.au
Arneis, Dolcetto, Meunier, Moscato, Tempranillo

Boyntons Feathertop (Alpine Valleys)
www.boynton.com.au
Durif, Pinot gris, Prosecco, Saperavi, Sangiovese, Savagnin, Tempranillo, Vermentino, Viognier

Bracken Hill (Southern Tasmania)
Gewurztraminer

Brairose Estate (Margaret River)
briarose.com.au
Cabernet franc

Brammar Estate (Yarra Valley)
www.brammarestatewinery.com.au
Gewurztraminer, Pinot gris, Verdelho, Viognier

Brandy Creek Wines (Gippsland)
www.brandycreekwines.com.au
3Pinot gris, Meunier, Tempranillo

Brave Goose Vineyard (Goulburn Valley)
www.bravegoosevineyard.com.au
Gamay, Malbec, Viognier

Bream Creek Vineyard (Southern Tasmania)
www.potterscroft.com.au
Gewurztraminer, Schonburger

Bremerton (Langhorne Creek)
www.bremerton.com.au
Chenin blanc, Malbec, Petit verdot, Verdelho

Briar Ridge Vineyard (Hunter Valley)
www.briarridge.com.au
Gewurztraminer, Verdelho

Briarose Estate (Margaret River)
www.briarose.com.au
Cabernet franc

Bridgeman Downs Cellars (South Burnett)
www.bridgemandowns.com
Verdelho

Brierley Wines (Queensland Zone)
www.brierleywines.com
Chambourcin

Briery Estate (Perth Hills)
www.brieryestatewines.com
Furmint, Grenache, Harslevelu, Verdelho, Zante

Brindabella Hills (Canberra)
brindabellahills.com.au
Sangiovese, Viognier

Brini Estate (McLaren Vale)
www.briniwines.com.au
Grenache

Broke's Promise (Hunter Valley)
www.brokespromise.com.au
Barbera, Verdelho

Broke Estate (Hunter Valley)
www.ryanwines.com.au
Barbera, Tempranillo

Broken Gate Wines (Heathcote)
www.brokengate.com.au
Sangiovese

Broken River Vineyards (Goulburn Valley)
www.brokenrivervineyards.com.au
Chenin blanc, Grenache, Mourvedre

Brokenwood Wines (Hunter Valley)
www.brokenwood.com.au
Pinot gris, Nebbiolo, Sangiovese, Viognier

Brookhampton Estate (Geographe)
www.brookhamptonestate.com.au
Grenache, Tempranillo, Viognier

Brookwood Estate (Margaret River)
www.brookwood.com.au
Chenin blanc

Broomstick Estate (Margaret River)
www.broomstick.com.au
Petit verdot

Brown's Farm Winery (Hunter Valley)
Pinot gris

Brown Brothers (King Valley)
www.brown-brothers.com.au
Aglianico, Arneis, Barbera, Carmenere, Chenin blanc, Cienna, Crouchen, Dolcetto, Durif, Flora, Graciano, Lagrein, Lexia, Mondeuse, Moscato, Muscadelle, Nebbiolo, Nero d'Avola, Petit verdot, Orange muscat, Roussanne, Sangiovese, Savagnin, Tarrango, Tempranillo, Vermentino, Viognier, Zibibbo, Zinfandel

Brown Magpie Wines (Geelong)
www.brownmagpiewines.com
Pinot gris

Browns of Padthaway (Padthaway)
www.browns-of-padthaway.com
Malbec, Verdelho

Brumby Wines (Swan Hill)
www.brumbywines.com.au
Durif

Brumfield (Yarra Valley)
www.brumfield.com.au
Marsanne

Brush Box Vineyard (Hunter Valley)
Verdelho

Buckshot Vineyard (Heathcote)
www.buckshotvineyard.com.au
Zinfandel

Bullers Beverford (Swan Hill)
www.buller.com.au
Chasselas, Chenin blanc, Cinsaut, Grenache, Moscato, Mourvedre

Bullers Calliope (Rutherglen)
www.buller.com.au
Durif, Marsanne, Mondeuse, Muscadelle, Tempranillo, Vermentino

Bulong Estate (Yarra Valley)
www.bulongestate.com
Cabernet franc, Pinot gris

Bungawarra (Granite Belt)
bungawarrawines.com.au
Gewurztraminer, Malbec

Burge Family Winemakers (Barossa Valley)
www.burgefamily.com.au
Grenache, Mourvedre, Souzao, Touriga, Zinfandel

Burke and Wills Winery (Heathcote)
www.wineandmusic.net
Gewurztraminer

Burnbrae Winery (Mudgee)
www.burnbraewines.com
Petit verdot, Sangiovese

By Farr (Geelong)
www.byfarr.com.au
Viognier

Byramgou Park (Geographe)
www.byramgou.com.au
Grenache

WINERIES • C 141

Byrne and Smith (McLaren Vale)
www.byrneandsmith.com.au
Cabernet franc, Tempranillo

C

Calais Estate (Hunter Valley)
www.calaiswines.com.au
Chambourcin, Durif, Marsanne, Verdelho, Viognier, Zinfandel

Calico Town (Rutherglen)
www.thewickedvirgin.com
Trebbiano

Callipari Wine (Murray Darling)
www.callipari.com
Grenache

Cambewarra Estate (Shoalhaven Coast)
www.cambewarraestate.com.au
Chambourcin, Verdelho

Camp Road Estate (Hunter Valley)
www.camproadestate.com.au
Shiraz Viognier, Verdelho, Viognier

Campbells Wines (Rutherglen)
www.campbellswines.com.au
Durif, Malbec, Muscadelle, Pedro Ximenez, Roussanne, Ruby Cabernet, Trebbiano, Viognier

Camyr Allyn Wines (Hunter Valley)
www.camyrallynwines.com.au
Verdelho

Canolobas-Smith (Orange)
www.clearviewwines.com.au
Chambourcin

Canonbah Bridge (Western Plains)
www.canonbah.com.au
Grenache, Mourvedre, Verdelho

Canungra Valley (Queensland Coastal)
www.canungravineyards.com.au
Chambourcin, Chenin blanc, Verdelho

Cape Banks (Limestone Coast Zone)
Pinot gris

Cape Barren Wines (McLaren Vale)
www.capebarrenwines.com.au
Grenache, Mourvedre

Cape Bernier Vineyard (Southern Tasmania)
www.capebernier.com.au
Pinot gris

Cape Bouvard (Peel)
Chenin blanc, Grenache, Verdelho

Cape Grace Wines (Margaret River)
www.capegracewines.com.au
Chenin blanc

Cape Horn Vineyard (Goulburn Valley)
www.capehornvineyard.com.au
Durif, Marsanne, Zinfandel

Cape Mentelle (Margaret River)
www.capementelle.com.au
Grenache, Marsanne, Mourvedre, Roussanne, Sangiovese, Viognier, Zinfandel

Capel Vale (Geographe)
www.capelvale.com
Chenin blanc, Nebbiolo, Sangiovese, Tempranillo, Verdelho, Viognier

Capercaillie (Hunter Valley)
www.capercailliewine.com.au
Chambourcin, Gewurztraminer, Petit verdot

Capital Wines (Canberra)
www.capitalwines.com.au
Tempranillo

Carabooda Estate (Swan District)
www.caraboodaestatewines.com.au
Chenin blanc

Cardinam Estate (Clare Valley)
www.cardinham.com
Malbec, Sangiovese

Cargo Road Wines (Orange)
www.cargoroadwines.com.au
Gewurztraminer, Zinfandel

Carickalinga Creek (Southern Fleurieu)
www.ccvineyard.com.au
Viognier

Carilley Estate (Swan Valley)
www.carilleyestate.com.au
Chenin blanc, Grenache, Malbec, Viognier

Carlaminda Estate (Geographe)
www.carlaminda.com
Chenin blanc, Nebbiolo, Tempranillo, Viognier

Carlei Estate (Yarra Valley)
www.carlei.com.au
Barbera, Nebbiolo, Sangiovese

Carn Estate (Murray Darling)
Colombard

Carosa (Perth Hills)
www.carosavineyard.com
Chenin blanc, Verdelho

Carpenteri Vineyards (Swan Hill)
Grenache, Malbec, Mourvedre

Casa Freschi (Langhorne Creek)
www.casafreschi.com.au
Malbec, Nebbiolo

Cascabel (McLaren Vale)
www.cascabelwinery.com.au
Cinsaut, Graciano, Grenache, Mourvedre, Roussanne, Tempranillo, Viognier

Casella (Riverina)
www.casellawine.com.au
Moscato, Petit verdot, Pinot gris, Sangiovese, Tempranillo, Verdelho, Viognier

Casley Mount Hutton Winery (Granite Belt)
www.casley.com.au
Chenin blanc

Cassegrain (Hastings River)
www.cassegrainwines.com.au
Chambourcin, Gewurztraminer, Verdelho

Castagna Vineyard (Beechworth)
www.castagna.com.au
Sangiovese, Viognier

Cathcart Ridge Estate (Grampians)
www.cathcartwines.com.au
Chasselas

Catherine's Ridge (Cowra)
Verdelho

Catherine Vale Vineyard (Hunter Valley)
www.catherinevale.com.au
Arneis, Barbera, Dolcetto, Verdelho

Catspaw Farm (Granite Belt)
www.catspawfarm.cjb.net
Barbera, Chambourcin, Roussanne, Sangiovese

Caught Redhanded (Adelaide Zone)
Pinot gris

Cedar Creek Estate (Queensland Coastal)
www.cedarcreekestate.com.au
Chambourcin, Verdelho

Celestial Bay (Margaret River)
www.celestialbay.com.au
Malbec, Petit verdot, Ruby Cabernet

Centennial Vineyards (Southern Highlands)
www.centennial.net.au
Barbera, Corvina, Meunier, Pinot gris, Rondinella, Savagnin, Tempranillo, Verdelho

Ceravolo Premium Wines (Adelaide Plains)
www.ceravolo.com.au
Pinot gris, Petit verdot, Sangiovese

Ceres Bridge Estate (Geelong)
Nebbiolo, Pinot gris, Tempranillo, Viognier

Chain of Ponds (Adelaide Hills)
www.chainofponds.com.au
Barbera, Grenache, Pinot gris, Nebbiolo, Sangiovese, Viognier

WINERIES • C 143

Chalice Bridge Estate (Margaret River)
www.chalicebridge.com.au
Viognier

Chalk Hill Winery (McLaren Vale)
www.chalkhill.com.au
Barbera, Cabernet franc, Moscato, Sangiovese

Chalmers (Murray Darling)
www.chalmerswine.com.au
Aglianico, Lagrein, Fiano, Greco bianco, Malvasia Istriana, Negro amaro, Nero d'Avola, Nosiola, Sangiovese, Sagrantino, Schioppettino, Tempranillo, Vermentino, Viognier

Chambers Rosewood (Rutherglen)
www.chambersrosewood.com.au
Cinsaut, Gouais blanc, Mondeuse, Muscadelle, Palomino, Roussanne

Channybearup (Pemberton)
www.channybearup.com.au
Verdelho

Chapel Hill (McLaren Vale)
www.chapelhillwine.com.au
Malbec, Pinot gris, Sangiovese, Savagnin, Tempranillo, Verdelho

Chaperon Wines (Bendigo)
www.chaperon.com.au
Grenache, Mourvedre

Chapman's Creek Vineyard (Margaret River)
Chenin blanc

Chapman Valley Wines (Central Western Australian Zone)
www.chapmanvalleywines.com.au
Chenin blanc, Verdelho, Zinfandel

Charlatan Wines (McLaren Vale)
Sangiovese

Charles Melton (Barossa Valley)
www.charlesmeltonwines.com.au
Grenache, Meunier, Mourvedre, Muscadelle, Pedro Ximenez

Charlies Estate Wines (Swan Valley)
www.charliesestatewines.com.au
Chenin blanc, Marsanne, Tempranillo, Viognier

Chartley Estate (Northern Tasmania)
chartleyestatevineyard.com.au
Pinot gris

Chateau Champsaur (Central Ranges Zone)
Colombard

Chateau Dorrien (Barossa Valley)
Grenache, Mourvedre, Trebbiano

Chateau Mildura (Murray Darling)
www.chateaumildura.com.au
Pinot gris, Petit verdot, Viognier

Chateau Tanunda (Barossa Valley)
www.chateautanunda.com
Moscato, Pinot gris, Zinfandel

Chatsfield (Mount Barker)
www.chatsfield.com.au
Cabernet franc, Gewurztraminer

Cheriton (Swan District)
www.cheriton.com.au
Chenin blanc

Chestnut Grove (Manjimup)
www.chestnutgrove.com.au
Verdelho

Chestnut Hill Vineyard (Port Phillip Zone)
www.chestnuthillvineyard.com.au
Nebbiolo

Chidlows Well (Central Western Australian Zone)
chidlowswell.com.au
Chenin blanc, Verdelho

Chislehurst Estate (Hunter Valley)
Verdelho

Chittering Valley Winery (Perth Hills)
Chenin blanc, Grenache, Pedro Ximenez, Zinfandel

Chrismont (King Valley)
www.chrismont.com.au
Arneis, Barbera, Marzemino, Petit manseng, Pinot gris, Prosecco, Sangiovese, Savagnin, Tempranillo

Churchview Estate (Margaret River)
www.churchview.com.au
Marsanne

Ciavarella (King Valley)
www.oxleyestate.com.au
Aucerot, Dolcetto, Durif, Graciano, Verdelho, Viognier

Ciccone Estate (King Valley)
www.cicconewines.com.au
Sangiovese

Circo V (King Valley)
www.cirko-v.com.au
Viognier

Cirillo (Barossa Valley)
Grenache

Clancy's of Conargo (Riverina)
www.clancy.com.au
Mourvedre, Taminga, Verdelho

Clancy Fuller (Barossa Valley)
www.clancyfuller.com.au
Mourvedre, Grenache

Clarence Hill (McLaren Vale)
www.clarencehillwines.com.au
Grenache

Clarendon Hills (McLaren Vale)
www.clarendonhills.com.au
Grenache

Classic McLaren Wines (McLaren Vale)
www.classicmclarenwines.com.au
Grenache

Claymore Wines (Clare Valley)
www.claymorewines.com.au
Grenache, Mourvedre, Viognier

Clearview Estate Mudgee (Mudgee)
www.clearviewwines.com.au
Barbera, Pinot gris, Sangiovese

Cleggett (Langhorne Creek)
www.cleggettwines.com.au
Malian, Shalistin

Cleveland (Macedon Ranges)
www.clevelandwinery.com.au
Pinot gris

Clonakilla (Canberra)
www.clonakilla.com.au
Viognier

Cloudbreak Wines (Adelaide Hills)
cloudbreakwines.com.au
Pinot gris

Clovely Estate (South Burnett)
www.clovely.com.au
Barbera, Petit verdot, Verdelho

Clyde Park Vineyard (Geelong)
www.clydepark.com.au
Pinot gris

Cobaw Ridge (Macedon Ranges)
www.cobawridge.com.au
Lagrein, Vermentino, Viognier

Cobbitty Wines (South Coast Zone)
Barbera, Grenache, Trebbiano

Cofield Wines (Rutherglen)
www.cofieldwines.com
Cabernet franc, Chenin blanc, Durif, Gamay, Malbec, Muscadelle, Sangiovese

Coldstream Hills (Yarra Valley)
www.coldstreamhills.com.au
Pinot gris

Collina del Re (King Valley)
Meunier

Colonial Estate (Barossa Valley)
www.colonialwine.com.au
Grenache, Mourvedre, Muscadelle

Colvin Wines (Hunter Valley)
colvinwines.com.au
Sangiovese

Concotton Creek (Peel)
Chenin blanc, Zinfandel

WINERIES • C 145

Connor Park (Bendigo)
www.connorparkwinery.com.au
Barbera, Durif, Marsanne, Mourvedre, Sangiovese

Constable Vineyards (Hunter Valley)
constablevineyards.com.au
Pinot gris, Verdelho

Conte Estate Wines (McLaren Vale)
www.conteestatewines.com.au
Gewurztraminer, Grenache, Viognier

Contessa Estate (McLaren Vale)
www.conteestatewines.com.au
Gewurztraminer, Grenache

Coolangatta Estate (Shoalhaven Coast)
www.coolangattaestate.com.au
Chambourcin, Tannat, Tempranillo, Verdelho

Coombe Farm Vineyard (Yarra Valley)
www.coombefarm.com.au
Arneis, Marsanne, Pinot gris, Viognier

Cooper wines (Hunter Valley)
www.cooperwines.com.au
Chambourcin, Verdelho

Copper Bull (Yarra Valley)
Sangiovese

Coriole (McLaren Vale)
www.coriole.com
Barbera, Chenin blanc, Fiano, Grenache, Nebbiolo, Sangiovese

Cosham (Perth Hills)
www.coshamwines.com
Petit verdot

Counterpoint Vineyard (Pyrenees)
www.counterpointvineyard.com.au
Malbec, Nebbiolo, Petit verdot, Roussanne, Sangiovese, Tempranillo, Viognier

Cow Hill (Beechworth)
www.cowhill.com.au
Nebbiolo, Tempranillo, Viognier

Crabtree of Watervale (Clare Valley)
www.crabtreewines.com.au
Muscadelle, Tempranillo, Zibibbo

Craigow (Southern Tasmania)
www.craigow.com.au
Gewurztraminer

Crane Wines (South Burnett)
www.cranewines.com.au
Chambourcin, Marsanne, Ruby Cabernet, Verdelho

Craneford (Barossa Valley)
www.cranefordwines.com
Grenache, Petit verdot, Viognier

Creed of Barossa (Barossa Valley)
www.creedwines.com
Viognier

Creeks Edge Wines (Mudgee)
www.creeksedge.com.au
Chambourcin, Verdelho

Crittenden at Dromana (Mornington Peninsula)
www.geppettowines.com
Arneis, Barbera, Dolcetto, Melon de Bourgogne, Moscato, Pinot gris, Sangiovese, Savagnin, Tempranillo

Crooked River Wines (Shoalhaven Coast)
www.crookedriverwines.com
Arneis, Chambourcin, Ruby Cabernet, Sangiovese, Verdelho

Cross Rivulet (Southern Tasmania)
Gewurztraminer, Muller Thurgau

Cruickshank Callatoota Wines (Hunter Valley)
www.cruickshank.com.au
Cabernet franc

Cullen Wines (Margaret River)
www.cullenwines.com.au
Malbec, Petit verdot

Cumulus Wines (Orange)
www.cumuluswines.com.au
Moscato, Pinot gris

Cupitt's Winery (Shoalhaven Coast)
www.cupittwines.com.au
Chambourcin, Viognier

Curlewis Winery (Geelong)
www.curlewiswinery.com.au
Pinot gris

Curly Flat (Macedon Ranges)
www.curlyflat.com
Pinot gris

Currans Family Wines (Murray Darling)
www.curransfamilywines.com.au
Carina, Durif, Grenache, Viognier

Currency Creek (Currency Creek)
www.currencycreekwines.com.au
Grenache

Cushendell (Southern Highlands)
cushendall.ajcmiskelly.id.au
Chenin blanc, Gamay

Cuttaway Hill Estate (Southern Highlands)
www.cuttawayhillwines.com.au
Pinot gris

Cynergie Wines (Goulburn Valley)
www.cynergiewines.com
Tarrango

Cypress Post (Granite Belt)
www.cypresspost.com.au
Marsanne, Viognier

D

D'Arenberg (McLaren Vale)
www.darenberg.com.au
Chambourcin, Cinsaut, Grenache, Marsanne, Mourvedre, Petit verdot, Roussanne, Sagrantino, Souzao, Tempranillo, Viognier

Dal Zotto Estate (King Valley)
dalzotto.com.au
Arneis, Barbera, Pinot gris, Prosecco, Sangiovese

Dalfaras (Nagambie Lakes)
www.tahbilk.com.au
Marsanne, Pinot gris, Sangiovese, Verdelho

Darling Estate (King Valley)
Chenin blanc, Gamay

Darling Park (Mornington Peninsula)
www.darlingparkwinery.com
Pinot gris, Viognier

Darlington Estate (Perth Hills)
www.darlingtonestate.com.au
Grenache

Date Brothers (Swan Hill)
www.datebroswines.com.au
Durif

David Hook Wines (Hunter Valley)
www.davidhookwines.com.au
Barbera, Pinot gris, Verdelho, Viognier

David Treager (Nagambie Lakes)
www.dromanaestate.com.au
Tempranillo, Verdelho, Viognier

Dawson and Wills (Strathbogie Ranges)
Tempranillo

Daylesford Wine Company (Macedon Ranges)
Refosco

Dayleswood Winery (Riverina)
Chambourcin

De Beaurepaire Wines (Mudgee)
www.debeaurepairewines.com
Petit verdot, Verdelho, Viognier

De Bortoli (Riverina)
www.debortoli.com.au
Arneis, Colombard, Moscato, Petit verdot, Verdelho, Zinfandel

De Iulius (Hunter Valley)
www.dewine.com.au
Verdelho

De Lisio Wines (McLaren Vale)
Grenache, Pinot gris

WINERIES • D

de Mestre Wines (Mudgee)
www.demestrewines.com.au
Viognier

Deakin Estate (Murray Darling)
www.deakinestate.com.au
Colombard, Petit verdot, Tannat, Tempranillo, Viognier

Deep Woods Estate (Margaret River)
www.deepwoods.com.au
Verdelho

Deisen (Barossa Valley)
www.deisen.com.au
Grenache, Mourvedre

Del Rios (Geelong)
www.delrios.com.au
Marsanne

Delatite Winery (Upper Goulburn)
www.delatitewinery.com.au
Gewurztraminer, Malbec, Nebbiolo, Pinot gris

Deviation Road (Adelaide Hills)
www.deviationroad.com
Pinot gris, Sangiovese

Di Fabio Estate (McLaren Vale)
www.difabioestatewines.com.au
Grenache, Mourvedre, Petit verdot

Di Lusso Estate (Mudgee)
www.dilusso.com.au
Aglianico, Aleatico, Barbera, Nebbiolo, Picolit, Pinot gris, Sangiovese, Vermentino

Diamond Valley Vineyards (Yarra Valley)
www.diamondvalley.com.au
Viognier

Diggers Bluff (Barossa Valley)
www.diggersbluff.com
Alicante Bouchet, Grenache, Mourvedre

Diloreto Wines (Adelaide Plains)
Grenache, Mourvedre

Dimbulla (Hunter Valley)
Tempranillo

Dinny Goonan Family Estate (Geelong)
www.dinnygoonan.com.au
Malbec

Disaster Bay Wines (South Coast Zone)
www.disasterbaywines.com
Malbec, Mourvedre, Petit verdot

Divers Luck Wines (Northern Rivers Zone)
Chambourcin, Verdelho

Djinta Djinta (Gippsland)
www.djintadjinta.com.au
Marsanne, Roussanne, Ruby Cabernet, Viognier

Doc Adams (McLaren Vale)
www.docadamswines.com.au
Pinot gris

Doctors Nose Wines (New England)
www.doctorsnosewines.com.au
Grenache, Petit verdot, Tempranillo, Verdelho

DogRidge (McLaren Vale)
www.dogridge.com.au
Grenache, Petit verdot, Viognier

Dogrock Winery (Pyrenees)
www.dogrock.com.au
Grenache, Tempranillo

Domain Barossa (Barossa Valley)
www.domainbarossa.com
Grenache, Mourvedre

Domain Day (Barossa Valley)
www.domaindaywines.com
Garganega, Lagrein, Sagrantino, Sangiovese, Saperavi, Viognier

Dominic Versace Wines (Adelaide Plains)
www.dominicversace.com.au
Grenache, Moscato, Sangiovese

Donegal Wines (Riverland)
Verdelho

Donnelly River Wine (Pemberton)
www.donnellyriverwines.com.au
Chenin blanc

Donnybrook Estate (Geographe)
Barbera, Cinsaut, Graciano, Grenache, Tempranillo, Verdelho, Zinfandel

Donovan Wines (Grampians)
www.donovanwines.com.au
Tempranillo

Dookie College Winery (Goulburn Valley)
www.dookie.unimelb.edu.au
Tarrango

Doonkuna Wines (Canberra)
www.doonkuna.com.au
Sangiovese

Dos Rios (Swan Hill)
www.dosrios.com.au
Durif, Moscato, Pinot gris, Tempranillo, Verdelho, Viognier

Douglas Vale (Hastings River)
www.douglasvalevineyard.com.au
Chambourcin, Fragola, Villard blanc

Dowie Doole (McLaren Vale)
www.dowiedoole.com
Chenin blanc, Petit verdot, Viognier

Drakesbrook Wines (Peel)
www.drakesbrookwines.com.au
Petit verdot

Draytons (Hunter Valley)
www.draytonswines.com.au
Verdelho

Drews Creek Wines (Hunter Valley)
Sangiovese

Drinkmoor Wines (Rutherglen)
www.drinkmoorwines.com
Durif, Chenin blanc, Petit verdot

Dromana Estate (Mornington Peninsula)
www.dromanaestate.com.au
Arneis, Barbera, Dolcetto, Nebbiolo, Pinot gris, Sangiovese

Duke's Vineyard (Porongurup)
www.dukesvineyard.com
Petit verdot

Dumaresq Valley Vineyard (New England)
www.dumaresqvalleyvineyard.com.au
Barbera, Tempranillo

Dunn's Creek Winery (Mornington Peninsula)
www.dunnscreek.com.au
Arneis, Barbera, Savagnin, Tempranillo

Dusty Hill Estate (South Burnett)
www.dustyhill.com.au
Verdelho

Dyson Wines (McLaren Vale)
www.dysonwines.com
Viognier

E

Eagle Vale (Margaret River)
www.eaglevalewine.com
Petit verdot

Eaglerange Wines (Alpine Valleys)
www.happyvalley75.com.au/eaglerange
Tempranillo, Viognier

Eagles Rise (Geelong)
wildwine.com.au
Pinot gris

Echo Ridge Wines (Hunter Valley)
www.echoridgewines.com.au
Verdelho

Eden Hall (Eden Valley)
www.edenhall.com.au
Viognier

Eden Road Wines (Eden Valley)
www.edenroadwines.com.au
Grenache

Edwards and Chaffey (McLaren Vale)
www.edwardsandchaffey.com.au
Sangiovese, Tempranillo

WINERIES • F   149

Elan Vineyard (Mornington Peninsula)
www.elanvineyard.com.au
Gamay

Elderton (Barossa Valley)
www.eldertonwines.com.au
Verdelho, Zinfandel

Eldredge (Clare Valley)
www.winediva.com.au
Malbec, Sangiovese

Eldridge Estate (Mornington Peninsula)
www.eldridge-estate.com.au
Gamay

Eleven Paddocks (Pyrenees)
www.elevenpaddocks.com.au
Petit verdot

Elgee Park (Mornington Peninsula)
www.elgeeparkwines.com.au
Pinot gris, Viognier

Eling Forest Winery (Southern Highlands)
www.elingforest.com.au
Furmint, Harslevelu

Ellender Estate (Macedon Ranges)
www.ellenderwines.com
Pinot gris

Elliot Rocke Estate (Mudgee)
www.elliotrockeestate.com.au
Doradillo, Gewurztraminer

Elsewhere Vineyard (Southern Tasmania)
www.elsewherevineyard.com
Gewurztraminer

Elsmore's Caprera Grove (Hunter Valley)
www.elsmorewines.com.au
Verdelho

Elysium Vineyard (Hunter Valley)
www.elysiumvineyardcottage.com.au
Verdelho

Emmas Cottage Vineyard (Hunter Valley)
www.emmascottage.com.au
Verdelho

Emmetts Crossing Wines (Peel)
Malbec, Verdelho

Eperosa (Barossa Valley)
www.eperosa.com.au
Grenache, Mourvedre

Epsilon (Barossa Valley)
www.epsilonwines.com.au
Graciano, Montepulciano, Tempranillo

Ermes Estate (Mornington Peninsula)
Malvasia, Pinot gris

Ernest Hill Wines (Hunter Valley)
www.ernesthillwines.com.au
Gewurztraminer, Verdelho

Eumundi Winery (Queensland Coastal)
www.eumundiwinery.com.au
Chambourcin, Durif, Mondeuse, Petit verdot, Regent, Savagnin, Sirius, Taminga, Tannat, Tempranillo, Verdelho

Evans and Tate (Margaret River)
www.evansandtate.com.au
Chenin blanc, Moscato, Verdelho

Evelyn County Estate (Yarra Valley)
www.evelyncountyestate.com.au
Tempranillo

Even Keel Wines (Canberra)
evenkeelwines.com.au
Shiraz Viognier

Eyre Creek (Clare Valley)
www.eyrecreekwines.com.au
Grenache

F

Faber Vineyard (Swan Valley)
www.fabervineyard.com.au
Malbec, Petit verdot, Verdelho

150 WINERIES • F

Fairview Wines (Hunter Valley)
www.fairviewwines.com.au
Barbera, Chambourcin, Verdelho

Farago Hill (Southern Highlands)
www.faragohill.com.au
Pinot gris

Faranda Wines (Swan District)
Chasselas, Grenache

Farosa Estate (Adelaide Plains)
Mourvedre

Farrell Estate (Murray Darling)
www.farrellestate.com.au
Sangiovese, Viognier

Farrells Limestone Creek (Hunter Valley)
www.farrelswines.com.au
Verdelho

Felsberg Winery (Granite Belt)
Gewurztraminer, Sylvaner

Ferguson Falls Winery (Geographe)
Tempranillo, Nebbiolo

Fermoy Estate (Margaret River)
www.fermoy.com.au
Chenin blanc, Nebbiolo, Verdelho

Fern Gully Winery (Shoalhaven Coast)
www.shoalhavencoast.com.au
Chambourcin

Fernbrook Estate (Porongurup)
Gamay

Fernfield Wines (Eden Valley)
www.fernfieldwines.com.au
Gewurztraminer

Ferngrove Vineyards (Frankland River)
www.ferngrove.com.au
Malbec

Fighting Gully Road (Beechworth)
Petit manseng, Sangiovese, Tempranillo, Viognier

Fireblock (Clare Valley)
Grenache

First Creek (Hunter Valley)
www.firstcreekwines.com.au
Verdelho, Viognier

First Drop (Barossa Valley)
www.firstdropwines.com
Arneis, Barbera, Montepulciano, Nebbiolo, Savagnin, Tinta amarela, Touriga

Fish Tail Wines (Swan Valley)
www.fishtailwines.com.au
Chenin blanc, Verdelho

Five Geese Hillgrove Wines (McLaren Vale)
www.fivegeese.com
Grenache

Five Sons Estate (Mornington Peninsula)
www.fivesons.com.au
Pinot gris

Flat View Vineyard (Clare Valley)
www.byrneandsmith.com.au
Grenache, Nebbiolo, Sangiovese

Flinders Bay (Margaret River)
www.flindersbaywines.com.au
Chenin blanc, Verdelho

Flying Duck Estate (King Valley)
www.flyingduckestate.com.au
Sangiovese, Viognier

Flying Fish Cove (Margaret River)
www.flyingfishcove.com
Chenin blanc, Nebbiolo, Sangiovese

Flynn's Wines (Heathcote)
www.flynnswines.com
Sangiovese, Verdelho, Viognier

Foate's Ridge (Hunter Valley)
www.foate.com.au
Verdelho

Foggo Wines (McLaren Vale)
www.foggowines.com.au
Cinsaut, Grenache, Viognier

Folino Estate (Alpine Valleys)
Fragola

Fonthill Wine (McLaren Vale)
Grenache, Verdelho

Fonty's Pool Vineyards (Pemberton)
www.fontyspool.com
Verdelho, Viognier

Fordwich Estate (Hunter Valley)
Verdelho

Forester Estate (Margaret River)
foresterestate.com.au
Alicante Bouchet

Foster e Rocco (Bendigo)
Sangiovese

Four Winds Vineyard (Canberra)
www.fourwindsvineyard.com.au
Sangiovese

Fox Creek Wines (McLaren Vale)
www.foxcreekwines.com
Grenache, Verdelho

Fox Gordon (Barossa Valley)
www.foxgordon.com.au
Fiano, Tempranillo, Viognier

Fox River Wines (Mount Barker)
www.foxriverwines.com.au
Chenin blanc

Foxey's Hangout (Mornington Peninsula)
www.foxeys-hangout.com.au
Moscato, Pinot gris, Vermentino

Franand Wines (Swan District)
Grenache

Francois Jacquard (Perth Hills)
Verdelho, Viognier

Frankland Estate (Frankland River)
www.franklandestate.com.au
Shiraz Viognier, Viognier

Freeman Vineyards (Hilltops)
www.freemanvineyards.com.au
Aleatico, Corvina, Pinot gris, Rondinella, Tempranillo, Viognier

French Island Vineyards (Mornington Peninsula)
www.fiv.com.au
Pinot gris

Freycinet (Southern Tasmania)
www.freycinetvineyard.com.au
Schonburger

Friday Creek Resort (Northern Rivers Zone)
www.fridaycreek.com
Chambourcin

Frog Rock (Mudgee)
www.frogrockwines.com
Chambourcin, Petit verdot, Pinot gris

Frogmore Creek Vineyard (Northern Tasmania)
frogmorecreekvineyards.com
Gewurztraminer, Pinot gris, Ruby Cabernet

Fryerstown Road Vineyard (Macedon Ranges)
www.fryerstownroadvineyard.com.au
Malbec

Fyffe Field (North East Victoria)
www.fyffefieldwines.com.au
Petit verdot, Touriga, Verdelho

G

Gabriel's Paddocks Vineyard (Hunter Valley)
www.gabrielspaddocks.com.au
Chenin blanc

Galafrey (Mount Barker)
galafreywines.com.au
Muller Thurgau

Galli Estate (Sunbury)
www.galliestate.com.au
Nebbiolo, Pinot gris, Sangiovese, Tempranillo, Viognier

WINERIES • G

Galvanized Wine Group (McLaren Vale)
Roussanne, Pinot gris, Roussanne, Viognier

Gapsted (Alpine Valleys)
www.gapstedwines.com.au
Barbera, Cabernet franc, Durif, Grenache, Moscato, Petit manseng, Pinot gris, Saperavi, Tarrango, Tempranillo, Touriga, Verdelho

Garbin Estate (Swan Valley)
Chenin blanc, Verdelho

Gartelmann Hunter (Hunter Valley)
www.gartelmann.com.au
Verdelho

Gawler River Grove (Adelaide Plains)
Grenache

Gecko Valley (Queensland Coastal)
www.geckovalley.com.au
Verdelho

Geddes Wines (McLaren Vale)
Grenache, Petit verdot, Viognier

Gehrig Estate (Rutherglen)
Durif, Muscadelle

Gelland Estate (Mudgee)
www.gellandestate.com.au
Viognier

Gemtree Vineyards (McLaren Vale)
www.gemtreevineyards.com.au
Grenache, Petit verdot, Sangiovese, Savagnin, Tempranillo, Viognier

Gentle Annie (Goulburn Valley)
www.gentle-annie.com
Verdelho

Geoff Hardy (McLaren Vale)
www.k1.com.au
Arneis, Grenache, Petit verdot

Geoff Merrill (McLaren Vale)
www.geoffmerrillwines.com
Grenache, Mourvedre, Viognier

Geographe Wines (Geographe)
www.geographewines.com.au
Nebbiolo, Sangiovese, Tempranillo

Gherardi Wines (Margaret River)
www.gherardi.com.au
Viognier

Ghost Riders Vineyard (Hunter Valley)
ghostriderswines.com.au
Viognier

Ghost Rock (Northern Tasmania)
www.ghostrock.com.au
Pinot gris

Giaconda (Beechworth)
https://www.giaconda.com.au
Roussanne

Giant Steps (Yarra Valley)
www.giant-steps.com.au
Petit verdot, Pinot gris, Viognier

Gibson Barossavale (Barossa Valley)
www.barossavale.com
Grenache, Mourvedre

Gilligan (McLaren Vale)
www.gilligan.com.au
Grenache, Marsanne, Mourvedre, Roussanne

Gin Gin Wines (Queensland Coastal)
Colombard, Gewurztraminer, Grenache, Malbec, Petit verdot, Sangiovese, Tempranillo, Verdelho

Gipsie Jack (Langhorne Creek)
www.gipsiejack.com
Malbec, Petit verdot

Glaetzer Wines (Barossa Valley)
www.glaetzer.com
Grenache

Glandore Estate (Hunter Valley)
www.glandorewines.com
Moscato, Savagnin, Tempranillo, Viognier

Glandore Estate (Hunter Valley)
www.glandorewines.com
Tempranillo, Verdelho

WINERIES • G

Glaymond Wines (Barossa Valley)
www.glaymondwines.com
Grenache, Tempranillo, Zinfandel

Glen Creek Wines (Upper Goulburn)
www.glencreekwines.com.au
Nebbiolo, Pinot gris

Glen Erin Vineyard Resort
(Macedon Ranges)
www.glenerinretreat.com.au
Gewurztraminer

Glendonbrook (Hunter Valley)
www.glendonbrook.com
Verdelho

Glenguin (Hunter Valley)
www.glenguinestate.com.au
Pinot gris, Sangiovese, Tannat

Glenwillow Vineyard (Bendigo)
www.glenwillow.com.au
Barbera, Nebbiolo

Gnadenfrei Estate (Barossa Valley)
www.treetopsbnb.com.au
Gewurztraminer, Grenache

Golden Ball (Beechworth)
www.goldenball.com.au
Grenache, Malbec

Golden Grape Estate (Hunter Valley)
www.goldengrape.com.au
Gewurztraminer

Golden Grove Estate (Granite Belt)
www.goldengrovee.com.au
Barbera, Durif, Tempranillo

Gomersal Wines (Barossa Valley)
www.gomersalwines.com
Grenache, Mourvedre

Gomersal Wines (Barossa Valley)
www.gomersalwines.com
Grenache, Mourvedre

Goombaronga Park (Alpine Valleys)
Sangiovese

Goona Warra Vineyard (Sunbury)
www.goonawarra.com.au
Cabernet franc, Roussanne

Goorambath (Glenrowan)
https://www.goorambath.com.au
Orange muscat, Pinot gris, Tannat, Verdelho

Goulburn Terrace (Nagambie Lakes)
www.goulburnterrace.com.au
Marsanne

Goundrey (Mount Barker)
www.goundreywines.com.au
Chenin blanc, Tempranillo, Verdelho

Governor's Choice Winery
(Queensland Zone)
www.governorschoice.com.au
Malbec, Verdelho

Gowrie Mountain Estate (Darling Downs)
www.gmewines.com.au
Chambourcin, Gamay, Tempranillo, Verdelho

Gracebrook Vineyards (King Valley)
www.gracebrook.com.au
Dolcetto, Moscato, Pinot gris, Sagrantino, Sangiovese, Savagnin

Grancari Estate (McLaren Vale)
www.grancariwines.com.au
Grenache

Grandview Vineyard (Southern Tasmania)
www.grandview.au.com
Gamay, Gewurztraminer

Granite Hills (Macedon Ranges)
www.granitehills.com.au
Tempranillo

Granite Ridge Wines (Granite Belt)
www.graniteridgewines.com.au
Petit verdot, Pinot gris, Ruby Cabernet, Tempranillo, Verdelho

Grant Burge (Barossa Valley)
www.grantburgewines.com.au
Grenache, Moscato, Mourvedre, Pinot gris, Ruby Cabernet, Viognier

154 WINERIES • H

Grassy Point Coatsworth Wines (Geelong)
www.grassypointwines.com.au
Cabernet franc, Malbec

Great Lakes Wines (Northern Rivers Zone)
www.greatlakeswines.com.au
Chambourcin, Verdelho

Green Valley Vineyard (Margaret River)
www.greenvalleyvineyard.com.au
Chenin blanc

Greenock Creek Wines (Barossa Valley)
Grenache

Greenstone Vineyard (Heathcote)
www.greenstoneofheathcote.com
Mourvedre, Sangiovese, Tempranillo

Grevillea Estate (South Coast Zone)
www.grevilleawines.com
Gewurztraminer

Grey Sands (Northern Tasmania)
www.greysands.com
Petit verdot, Pinot gris, Touriga, Viognier

Grove Estate Wines (Hilltops)
www.groveestate.com.au
Barbera, Nebbiolo, Petit verdot, Sangiovese, Viognier, Zinfandel

Growlers Gully (Upper Goulburn)
www.growlersgully.com.au
Marsanne, Roussanne

GrumbleBone Estate Vineyard (Southern Highlands)
Pinot gris

Gundowringla Wines (Alpine Valleys)
Malbec, Viognier

H

Haan (Barossa Valley)
www.haanwines.com.au
Viognier

Hackersley (Geographe)
www.hackersley.com.au
Mondeuse, Petit verdot, Verdelho

Hahndorf Hill (Adelaide Hills)
www.hahndorfhillwinery.com.au
Lemberger, Pinot gris, Trollinger

Hainault (Perth Hills)
www.hainault.com.au
Gewurztraminer, Sylvaner

Halifax (McLaren Vale)
www.halifaxwines.com.au
Vermentino

Halina Brook (Central Western Australian Zone)
Chenin blanc, Grenache, Verdelho

Hamiltons Bluff (Cowra)
www.hamiltonsbluff.com.au
Sangiovese, Viognier

Hamiltons Ewell Vineyards (Barossa Valley)
www.hamiltonewell.com.au
Grenache, Mourvedre

Hanging Rock Winery (Macedon Ranges)
www.hangingrock.com.au
Grenache, Marsanne, Mourvedre, Petit verdot, Pinot gris, Zinfandel

Hankin Estate (Goulburn Valley)
Verdelho

Happs (Margaret River)
www.happs.com.au
Bastardo, Cabernet franc, Carignan, Chenin blanc, Cinsaut, Furmint, Gamay, Graciano, Marsanne, Mourvedre, Muscadelle, Nebbiolo, Tinto Cao, Tempranillo, Verdelho, Viognier

WINERIES • H 155

Harcourt Valley (Bendigo)
www.harcourtvalley.com.au
Cabernet franc, Malbec, Meunier

Hare's Chase (Barossa Valley)
www.hareschase.com.au
Tempranillo

Harkaway Estate (Murray Darling)
Petit verdot

Harlow Park Estate (Mornington Peninsula)
Pinot gris

Harman's Ridge Estate (Margaret River)
www.harmansridge.com.au
Grenache, Marsanne

Harrington Glen Estate (Granite Belt)
Verdelho

Harris Organic Wines (Swan Valley)
www.harrisorganicwine.com
Bastardo, Chenin blanc, Madeline Angevine, Muscadelle, Pedro Ximenez, Verdelho

Harris River Estate (Geographe)
www.harrisriverestate.com.au
Verdelho, Viognier

Hartley Estate (Perth Hills)
www.hartleyestate.com.au
Viognier

Hartz Barn Wines (Eden Valley)
www.hartzbarnwines.com.au
Lagrein

Haselgrove (McLaren Vale)
www.haselgrove.com.au
Grenache, Pinot gris, Viognier

Hastwell and Lightfoot (McLaren Vale)
www.hastwellandlightfoot.com.au
Cabernet franc, Tempranillo, Viognier

Hawkers Gate (McLaren Vale)
www.hawkersgate.com.au
Grenache, Saperavi

Hay Shed Hill Wines (Margaret River)
www.hatshedhill.com.au
Cabernet franc, Tempranillo

Haywards of Locksley (Strathbogie Ranges)
Petit verdot, Viognier

Hazyblur Wines (Kangaroo Island)
www.hazyblur.com
Pinot gris

Heafod Glen Winery (Swan Valley)
www.heafordglen.com.au
Chenin blanc, Verdelho, Viognier

Heartland Vineyard (Hunter Valley)
www.heartlandvineyard.com
Barbera, Verdelho, Viognier

Heartland Wines (Limestone Coast Zone)
www.heartlandwines.com.au
Dolcetto, Lagrein, Pinot gris, Sangiovese, Verdelho, Viognier

Heathcote Estate (Heathcote)
www.heathcoteestate.com
Grenache

Heathcote II (Heathcote)
www.heathcote2.com
Tempranillo

Heathcote Winery (Heathcote)
www.heathcotewinery.com.au
Marsanne, Tempranillo, Viognier

Heathvale (Eden Valley)
www.heathvalewines.com.au
Sagrantino

Heggies Vineyard (Eden Valley)
www.heggiesvineyard.com
Viognier

Heidenriech Estate (Barossa Valley)
www.heidenreichvineyards.com.au
Viognier

Helen's Hill Estate (Yarra Valley)
www.helenshill.com.au
Arneis, Viognier

Henderson Hardie (King Valley)
Gewurztraminer, Meunier, Pinot gris

Henley Hill (Yarra Valley)
www.henleyhillwines.com.au
Pinot gris, Viognier

Henry's Drive (Padthaway)
www.henrysdrive.com
Verdelho

Henry Holmes Wines (Barossa Valley)
www.woodbridgefarm.com
Grenache

Henschke (Eden Valley)
www.henschke.com.au
Gewurztraminer, Grenache, Mourvedre, Pinot gris, Viognier

Hently Farm Wines (Barossa Valley)
www.hentleyfarm.com.au
Grenache, Viognier, Zinfandel

Herbert Vineyard (Limestone Coast Zone)
www.herbertvineyard.com.au
Pinot gris

Heritage Estate (Granite Belt)
www.heritagewines.com.au
Cabernet franc, Durif, Verdelho

Heritage Farm (Goulburn Valley)
Gewurztraminer

Heron Lake Estate (Margaret River)
www.heronlake.com.au
Verdelho

Herons Rise (Southern Tasmania)
www.heronsrise.com.au
Muller Thurgau

Hewitson (Barossa Valley)
www.hewitson.com.au
Grenache, Mourvedre, Muscadelle, Tempranillo

Heytesbury Ridge (Geelong)
www.heytesburyridge.com.au
Pinot gris

Hickinbotham (Mornington Peninsula)
www.hickinbotham.biz
Aligote, Grenache, Taminga

Hidden Creek (Granite Belt)
www.hiddencreek.com.au
Gewurztraminer, Marsanne, Nebbiolo, Tempranillo, Verdelho

Higher Plane (Margaret River)
www.higherplanewines.com.au
Malbec, Viognier

Hills View (McLaren Vale)
www.hillsview.com.au
Verdelho, Malbec

Hillside Estate (Hunter Valley)
Gewurztraminer, Verdelho

Hillstowe (Adelaide Hills)
www.hillstowe.com.au
Pinot gris

Hillwood Vineyard (Northern Tasmania)
Pinot gris

Hobbs of Barossa Ranges (Barossa Valley)
www.hobbsvintners.com.au
Grenache, Viognier

Hoddles Creek Estate (Yarra Valley)
www.hoddlescreekestate.com.au
Pinot blanc, Pinot gris

Hollick Wines (Coonawarra)
www.hollick.com
Barbera, Saint Macaire, Sangiovese, Savagnin, Tempranillo

Hollyclare (Hunter Valley)
www.hollyclare.com.au
Aleatico

Home Hill (Southern Tasmania)
www.homehillwines.com.au
Sylvaner

Honey Moon Vineyard (Adelaide Hills)
www.honeymoonvineyard.com.au
Marsanne, Tempranillo, Viognier

Honeytree Estate (Hunter Valley)
www.honeytreewines.com
Clairette

Hope Estate (Hunter Valley)
www.hopeestate.com.au
Verdelho

Hopwood Winery (Goulburn Valley)
www.hopwood.com.au
Sangiovese, Verdelho

Hotham Ridge Winery (Peel)
www.hothamridge.com.au
Chenin blanc, Tempranillo, Viognier, Zinfandel

Houghton (Swan Valley)
www.houghton-wines.com
Chenin blanc, Tempranillo, Verdelho

House of Certain Views (Hunter Valley)
www.margan.com.au
Barbera, Viognier

Howard Park Wines (Margaret River)
Carnelian

Howard Vineyard (Adelaide Hills)
www.howardvineyard.com
Cabernet franc, Viognier

HPR Wines (Mornington Peninsula)
Pinot gris

Hudson's Peake Wines (Hunter Valley)
Verdelho

Hugh Hamilton (McLaren Vale)
www.hughhamiltonwines.com.au
Petit verdot, Sangiovese, Saperavi, Tempranillo, Verdelho, Viognier

Humphries Estate (Shoalhaven Coast)
Chambourcin

Hungerford Hill (Hunter Valley)
www.hungerfordhill.com.au
Pinot gris

Hunting Lodge Estate (South Burnett)
www.huntinglodgeestate.com.au
Gewurztraminer, Verdelho

Huntleigh Vineyards (Heathcote)
www.heathcotewinegrowers.com.au
Gewurztraminer

Hutton Vale (Eden Valley)
http://www.huttonvale.com
Grenache, Mourvedre

Ibis Wines (Orange)
www.ibiswines.com.au
Cabernet franc, Pinot gris

I

Idle Hands Wines (Hunter Valley)
www.kv.com.au
Verdelho

Idlewild (Hunter Valley)
www.wildbrokewines.com.au
Barbera, Cabernet franc

Idylwild Wines (Geographe)
Chenin blanc

Illalangi Wines (Riverland)
www.illalangi.com.au
Petit verdot, Verdelho

Immerse (Yarra Valley)
www.immerse.com.au
Shiraz Viognier

Indigo Wine Company (Beechworth)
www.indigovineyard.com.au
Nebbiolo, Pinot gris, Petit verdot, Roussanne, Sangiovese, Viognier

Inghams Skilly Ridge Wines (Clare Valley)
Tempranillo, Zinfandel

Inkwell (McLaren Vale)
www.inkwellwines.com
Viognier, Zinfandel

Inlam Estate (Northern Rivers Zone)
www.ilnam.com.au
Chambourcin

Inneslake (Hastings River)
www.inneslake.com.au
Pinot gris, Verdelho

Innocent Bystander (Yarra Valley)
www.iinocentbystander.com.au
Moscato, Pinot gris, Sangiovese, Viognier

Iron Gate Estate (Hunter Valley)
www.iron-gate-estate.com.au
Verdelho

Iron Pot Bay Wines (Northern Tasmania)
www.ironpotbay.com.au
Gewurztraminer, Pinot gris

Ironbark Hill Estate (Hunter Valley)
www.ironbarkhill.com.au
Verdelho

Irvine (Eden Valley)
www.irvinewines.com.au
Petit Meslier, Pinot gris, Savagnin, Tannat, Zinfandel

Island Brook (Margaret River)
www.islandbrook.com.au
Verdelho

Ivanhoe Wines (Hunter Valley)
www.ivanhoewines.com.au
Chambourcin, Gewurztraminer, Verdelho

Izway Wines (Barossa Valley)
Grenache, Mourvedre, Viognier

J

Jaengenya Wines (Goulburn Valley)
www.jaengenya.com.au
Colombard, Durif

Jamabro Wines (Barossa Valley)
www.jamabro.com.au
Grenache, Viognier

James Estate (Hunter Valley)
www.jamesestatewines.com.au
Sylvaner, Verdelho

Jamieson Estate (Mudgee)
www.jamiesonestate.com.au
Barbera, Petit verdot

Jamsheed (Yarra Valley)
www.jamsheed.com.au
Gewurztraminer

Jane Brook Estate (Swan Valley)
www.janebrook.com.au
Chenin blanc, Verdelho

Jarrah Ridge Winery (Perth Hills)
www.jarrahridge.com.au
Chenin blanc, Verdelho, Viognier, Zinfandel

Jarrets of Orange (Orange)
Marsanne, Verdelho

Jarvis Estate (Margaret River)
www.jarvisestate.com.au
Cabernet franc

Jasper Hill (Heathcote)
www.jasperhill.com
Grenache, Nebbiolo

Jasper Valley (Shoalhaven Coast)
www.jaspervalleywines.com.au
Verdelho

jb Wines (Barossa Valley)
www.jbwines.com
Clairette, Pinot blanc, Zinfandel

Jeanneret Wines (Clare Valley)
www.ascl.com/j-wines
Grenache, Moscato

Jeir Creek (Canberra)
www.jeircreekwines.com.au
Viognier

Jenke Vineyards (Barossa Valley)
www.jenkevineyards.com
Cabernet franc, Grenache

Jerusalem Hollow (Margaret River)
www.jerusalemhollowwines.com.au
Roussanne

Jester Hill Wines (Granite Belt)
www.jesterhillwines.com.au
Tempranillo, Verdelho

Jillian Wines (Grampians)
www.jillianwines.com.au
Crouchen

Jimbour Wines (Queensland Zone)
www.jimbour.com
Grenache, Petit verdot, Verdelho, Viognier

Jindalee Estate (Geelong)
www.jindaleewines.com.au
Gewurztraminer, Moscato

Jingalla (Great Southern)
www.jingallawines.com.au
Verdelho

Jinglers Creek (Northern Tasmania)
www.jinglerscreekvineyard.com.au
Pinot gris

Jinks Creek Winery (Gippsland)
www.jinkscreekwinery.com.au
Marsanne, Pinot gris, Sangiovese

Joadja Vineyards (Southern Highlands)
www.joadja.com
Malbec, Sangiovese

John Duval Wines (Barossa Valley)
www.johnduvalwines.com
Grenache, Mourvedre

John Gehrig Wines (King Valley)
www.johngehrigwines.com.au
Chenin blanc, Durif, Gamay, Meunier

John Kosovich Wines (Swan Valley)
www.johnkosovichwines.com.au
Verdelho

Jones Road (Mornington Peninsula)
www.jonesroad.com.au
Pinot gris

Jones Winery and Vineyard (Rutherglen)
www.joneswinery.com
Marsanne, Pedro Ximenez

Josef Chromy Wines (Northern Tasmania)
www.josefchromy.com.au
Gewurztraminer

Judds Warby Range Estate (Glenrowan)
www.warbyrange-estate.com.au
Durif

Juniper Estate (Margaret River)
www.juniperestate.com.au
Chenin blanc, Verdelho

Jupiter Creek Winery (Adelaide Hills)
Grenache

Juul Wines (Hunter Valley)
www.roseglenestate.com
Verdelho

Jylland Vineyard (Central Western Australian Zone)
www.jylland.com.au
Carnelian, Chenin blanc, Sangiovese, Verdelho

K

Kabminye Wines (Barossa Valley)
www.kabminye.com
Black frontignac, Carignan, Cinsaut, Grenache, Mourvedre, Roussanne, Zinfandel

Kaesler (Barossa Valley)
www.kaesler.com.au
Grenache, Mourvedre, Palomino, Tempranillo, Touriga, Viognier

Kahlon Estate Wines (Riverland)
www.kahlonestatewine.com.au
Petit verdot

Kalari Wines (Cowra)
www.kalariwines.com.au
Verdelho

Kalgan River Wines (Albany)
www.kalganriverwines.com.au
Viognier

Kalleske Wines (Barossa Valley)
www.kalleske.com
Chenin blanc, Grenache, Mourvedre, Petit verdot, Viognier

Kamberra (Canberra)
www.kamberra.com.au
Viognier

Kancoona Valley Wines (Alpine Valleys)
www.kancoonavalleywines.com.au
Verdelho

Kangarilla Road (McLaren Vale)
www.kangarillaroad.com
Grenache, Sangiovese, Tinta Negra Mole, Viognier, Zinfandel

Kangderaar Vineyard (Bendigo)
www.bendigowine.org.au
Gewurztraminer, Touriga

Karanto Vineyards (Langhorne Creek)
www.karanto.com.au
Aglianico, Fiano, Greco bianco, Lagrein, Pinot gris, Zinfandel

Karatta Wine (Limestone Coast Zone)
www.karattawines.com.au
Malbec

Karri Grove Estate (Margaret River)
Verdelho

Katoa Wines (Heathcote)
Sangiovese, Viognier

Kay Bros Amery (McLaren Vale)
www.kaybrothersamerywines.com
Grenache, Mourvedre, Viognier

Keith Tulloch Wine (Hunter Valley)
www.keithtullochwine.com.au
Petit verdot, Verdelho, Viognier

Kellermeister Wines (Barossa Valley)
kellermeister.com.au
Carignan, Gewurztraminer, Grenache, Mourvedre, Pedro Ximenez, Pinot gris, Savagnin, Sylvaner, Tempranillo

Kellybrook (Yarra Valley)
www.kellybrookwinery.dining.com.au
Gewurztraminer

Kennedy (Heathcote)
www.kennedyvintners.com.au
Mourvedre, Tempranillo

Kenton Hill (Adelaide Hills)
Nebbiolo

Kevin Sobels Wines (Hunter Valley)
www.sobelswines.com.au
Gewurztraminer, Petit verdot, Sangiovese, Verdelho

Kidman Coonawarra Wines (Coonawarra)
www.kcwines.com
Viognier

Kies Family (Barossa Valley)
www.kieswines.com.au
Bastardo

Kilgour Estate (Geelong)
www.kilgourestate.com.au
Pinot gris

Kilikanoon (Clare Valley)
www.kilikanoon.com.au
Grenache, Mourvedre, Viognier

Killara Estate (Yarra Valley)
www.killarapark.com.au
Pinot gris, Viognier

Kimber Wines (McLaren Vale)
www.kimberwines.com
Petit verdot

WINERIES • K

Kindred Spirit Wines (Strathbogie Ranges)
Viognier

King River Estate (King Valley)
www.kingriverestate.com.au
Barbera, Lagrein, Sangiovese, Verdelho, Vermentino, Viognier

Kingsdale Wines (Southern New South Wales Zone)
www.kinsdale.com.au
Malbec

Kingsley Grove (South Burnett)
www.kingsleygrove.com
Chambourcin, Sangiovese, Verdelho

Kingston Estate (Riverland)
www.kingstonestatewines.com
Arneis, Barbera, Durif, Petit verdot, Tempranillo, Verdelho, Viognier, Zinfandel

Kinloch Wines (Upper Goulburn)
www.kinlochwines.com.au
Meunier, Tempranillo

Kirkham Estate (Sydney Basin)
Gewurztraminer, Petit verdot

Kirrihill Estates (Clare Valley)
www.kirrihillwines.com.au
Grenache, Mourvedre

Kirrihill (Adelaide Hills)
www.www.kirrihillwines.com.au
Grenache, Mourvedre, Tempranillo, Pinot gris, Viognier

Kitty Crawford Estate (New England)
www.sssbbq.com.au/kitty_crawford_Estate
Grenache, Petit verdot

Kladis Estate (Shoalhaven Coast)
www.kladisestatewines.com.au
Grenache, Gewurztraminer, Muscadelle, Verdelho

Knappstein Wines (Clare Valley)
www.knappsteinwines.com.au
Gewurztraminer, Pinot gris

Kneedeep (Margaret River)
www.kneedeepwines.com.au
Chenin blanc

Knots Wines (Heathcote)
www.thebridgevineyard.com.au
Sangiovese

Knotting Hill Vineyard (Margaret River)
www.knottinghill.com.au
Verdelho, Viognier

Koltz (McLaren Vale)
www.koltzwines.com.au
Grenache, Mourvedre, Sangiovese, Viognier

Kominos Wines (Granite Belt)
www.kominoswines.com
Chenin blanc

Kongwak Hills Winery (Gippsland)
Malbec

Kooroomba (Queensland Zone)
Marsanne, Verdelho

Koppamura Wines (Wrattonbully)
www.koppamurrawines.com.au
Petit verdot

Kotai Estate (Geographe)
www.kotai.com.au
Cabernet franc, Chenin blanc

Kouark (Gippsland)
www.gourmetgippsland.com.au
Pinot gris, Viognier

Kraanwood (Southern Tasmania)
Schonburger

Kreglinger Estate (Mount Benson)
www.kreglinger.com.au
Gewurztraminer, Pinot gris

Krinklewood (Hunter Valley)
www.krinklewood.com
Tempranillo, Verdelho

Kulkunbulla (Hunter Valley)
www.kulkunbulla.com.au
Petit verdot, Tempranillo, Verdelho

Kurabana (Geelong)
www.kurabana.com
Pinot gris

Kurrajong Downs (New England)
www.kurrajongdownswines.com
Taminga

Kurtz Family Vineyards (Barossa Valley)
www.kurtzfamilyvineyards.com.au
Mourvedre, Petit verdot

Kyotmunga Estate (Perth Hills)
www.kyotmunga.com.au
Barbera, Chenin blanc, Grenache, Taminga

L

La Cantina King Valley (King Valley)
lacantinakingvalley.com.au
Barbera, Dolcetto, Nebbiolo, Sangiovese

La Curio (McLaren Vale)
www.lacuriowines.com
Grenache

La Linea (Adelaide Hills)
www.lalinea.com
Tempranillo

Ladbroke Grove (Coonawarra)
www.ladbrokegrove.com.au
Viognier

Lady Bay Winery (Southern Fleurieu)
www.ladybay.com.au
Pinot gris, Viognier

Lake Breeze (Langhorne Creek)
www.lakebreeze.com.au
Grenache, Malbec, Moscato, Petit verdot

Lake Charlotte Wines (Perth Hills)
Verdelho

Lake Moodemere (Rutherglen)
www.moodemerewines.com.au
Biancone, Cinsaut, Durif

Lambert Vineyards (Canberra)
www.lambertvineyards.com.au
Gewurztraminer, Pinot gris

Lamonts (Central Western Australian Zone)
www.lamonts.com.au
Chenin blanc, Pedro Ximenez, Verdelho, Viognier

Lancaster Wines (Swan Valley)
www.lancasterwines.com.au
Chenin blanc, Verdelho

Lancefield Winery (Macedon Ranges)
www.wineandmusic.net
Gewurztraminer

Landhaus Estate (Barossa Valley)
landhauswines.com.au
Grenache, Mourvedre

Lane's End Vineyard (Macedon Ranges)
www.lanesend.com.au
Cabernet franc

Langanook Wines (Bendigo)
langanookwines.com.au
Viognier

Langmeil (Barossa Valley)
www.langmeilwinery.com.au
Grenache, Mourvedre, Ondenc, Viognier

Lankeys Creek Wines (Tumbarumba)
www.lankeyscreekwines.com.au
Petit verdot

Lanzthomson Wines (Barossa Valley)
www.lanzthomson.com
Grenache, Mourvedre, Viognier

Lark Hill Winery (Canberra)
www.larkhillwine.com.au
Gruner Veltliner, Pinot gris, Viognier

WINERIES • L

Lashmar (Kangaroo Island)
www.lashmarwines.com
Nebbiolo, Viognier

Laughing Jack (Barossa Valley)
www.laughingjackwines.com
Grenache

Laurellyn Wines (New England)
www.laurellynwines.tenterfield.biz
Pinot gris, Verdelho

Lawson Hill (Mudgee)
Gamay, Gewurztraminer, Verdelho

Lazy Ballerina (McLaren Vale)
www.lazyballerina.com
Shiraz Viognier

Lazy River Estate (Western Plains)
www.lazyriverestate.com.au
Petit verdot

Lazzar Wines (Mornington Peninsula)
www.lazzarwines.com
Arneis, Lagrein, Pinot gris, Verduzzo

Leabrook Estate (Adelaide Hills)
www.leabrookestate.com
Cabernet franc, Gewurztraminer, Pinot gris

Leconfield (Coonawarra)
www.leconfieldwines.com
Petit verdot

Lerida Estate (Canberra)
www.leridaestate.com
Pinot gris, Shiraz Viognier, Viognier

Lethbridge Wines (Geelong)
www.lethbridgewines.com
Sangiovese, Viognier, Zinfandel

Leura Park Estate (Geelong)
www.leuraparkestate.com.au
Pinot gris

Liebichwein (Barossa Valley)
www.liebichwein.com.au
Petit verdot, Tempranillo

Lighthouse Peak (Tumbarumba)
Verdelho

Lilac Hill Estate (Swan Valley)
www.lilachillestate.com.au
Chenin blanc, Verdelho, Zinfandel

Lillian (Pemberton)
Graciano, Marsanne, Mourvedre, Roussanne, Viognier

Lilliput Wines (Rutherglen)
www.lilliputwinesofrutherglen.com
Petit verdot, Viognier

Lillydale Estate (Yarra Valley)
www.mcwilliams.com.au
Gewurztraminer

Lillypilly Estate (Riverina)
www.lillypilly.com
Gewurztraminer, Lexia, Moscato, Petit verdot, Tempranillo

Lilyvale Wines (Darling Downs)
www.lilyvalewines.com
Verdelho

Limb Vineyards (Barossa Valley)
Mourvedre

Linda Domas Wines (McLaren Vale)
www.ldwines.com.au
Grenache, Viognier

Lindenderry (Mornington Peninsula)
www.lindenderry.com.au
Pinot blanc

Lindenton Wines (Heathcote)
Marsanne, Verdelho, Viognier

Lindrum (Langhorne Creek)
www.lindrum.com
Verdelho

Linfield Road Wines (Barossa Valley)
www.annandalevineyards.com.au
Grenache

Lion Mill Winery (Perth Hills)
www.lionmillwinery.com
Zinfandel

Little's Winery (Hunter Valley)
www.littleswinery.com.au
Marsanne, Verdelho, Viognier

Little Bridge (Canberra)
www.littlebridgewines.com.au
Gamay, Grenache, Sangiovese

Little River Wines (Swan Valley)
www.littleriverwinery.com
Chenin blanc, Cinsaut, Marsanne, Viognier

Little Wine Company (Hunter Valley)
www.olivine.com.au
Gewurztraminer, Petit verdot, Pinot gris, Sangiovese, Tempranillo, Verdelho, Viognier

Llangibby Estate (Adelaide Hills)
www.llangibbyestate.com
Pinot gris, Tempranillo

Loan Wines (Barossa Valley)
www.loanwines.com.au
Alicante Bouchet

Loch Luna (Riverland)
Verdelho

Lochmoore (Gippsland)
Pinot gris

Logan Wines (Mudgee)
www.loganwines.com
Gewurztraminer, Moscato, Pinot gris, Viognier

Long Point Vineyard (Hastings River)
www.longpointvineyard.com.au
Chambourcin, Gewurztraminer

Long Rail Gully (Canberra)
www.longrailgully.com.au
Pinot gris

Longview Creek (Sunbury)
www.longviewcreek.com
Cabernet franc, Chenin blanc

Longview Vineyard (Adelaide Hills)
www.longviewvineyard.com.au
Nebbiolo, Viognier, Zinfandel

Lost Valley (Upper Goulburn)
www.lostvalleywinery.com
Cortese, Verdelho

Lou Miranda Estate (Barossa Valley)
www.loumirandaestate.com.au
Grenache, Mourvedre, Pinot gris

Louee Wines (Mudgee)
www.louee.com.au
Cabernet franc, Nebbiolo, Petit verdot, Pinot gris, Verdelho, Viognier

Lowe Family Wines (Mudgee)
www.lowewine.com.au
Barbera, Pinot gris, Roussanne, Sangiovese, Zinfandel

Lucas Estate (Granite Belt)
www.lucasestate.com.au
Tempranillo, Verdelho

Lucy's Run (Hunter Valley)
www.lucysrun.com
Verdelho

Luke Lambert Wines (Heathcote)
www.lukelambertwines.com.au
Nebbiolo

Lyre Bird Hill (Gippsland)
www.lyrebirdhill.com.au
Gewurztraminer

Lyrebird Ridge Organic Winery (South Coast Zone)
www.lyrebirdridge.com
Chambourcin

M

M. Chapoutier Australia (Mount Benson)
www.chapoutier.com
Marsanne, Viognier

Mabrook Estate (Hunter Valley)
www.mabrookestate.com
Verdelho

Mac Forbes Wines (Yarra Valley)
www.macforbeswines.com
Arneis, Barbera, Lemberger

WINERIES • M 165

Macaw Creek Wine (Mount Lofty Ranges Zone)
www.macawcreekwines.com.au
Grenache, Pedro Ximenez

Macketh House Historic Vineyard (Pyrenees)
www.mackerethhouse.com.au
Tempranillo

Macquarie Grove Vineyards (Western Plains)
Barbera, Carmenere, Sangiovese, Tempranillo, Viognier

Mad Dog Wines (Barossa Valley)
www.maddogwines.com
Sangiovese

Maddens Rise (Yarra Valley)
Viognier

Madew Wines (Canberra)
Pinot gris

MadFish Wines (Margaret River)
www.madfishwines.com.au
Sangiovese, Tempranillo

Madigan Vineyard (Hunter Valley)
www.madiganvineyard.com.au
Verdelho

Maglieri (McLaren Vale)
www.beringerblass.com.au
Barbera, Nebbiolo, Sangiovese

Magpie Estate (Barossa Valley)
Grenache, Mourvedre

Maleny Mountain Wines (Queensland Coastal)
www.malenymountainwines.com.au
Chambourcin, Durif, Verdelho

Mallee Estate (Riverland)
www.malleeestatewines.com.au
Tempranillo

Mandalay Road (Geographe)
www.mandalayestate.com.au
Zinfandel

Mann (Swan Valley)
Cygne blanc

Mansfield Wines (Mudgee)
www.mansfieldwines.com.au
Grenache, Mourvedre, Pedro Ximenez, Petit manseng, Savagnin, Tempranillo, Tinto Cao, Touriga, Vermentino, Zinfandel

Manton's Creek Vineyard (Mornington Peninsula)
www.mantonscreekvineyard.com.au
Gewurztraminer, Meunier, Pinot gris, Tempranillo

Mardia Wines (Barossa Valley)
Cabernet franc, Petit verdot

Margan Family (Hunter Valley)
www.margan.com.au
Barbera, Verdelho, Viognier

Marienberg (McLaren Vale)
Grenache, Mourvedre

Marions Vineyard (Northern Tasmania)
www.marionsvineyard.com
Muller Thurgau, Pinot gris

Maritime Estate (Mornington Peninsula)
Pinot gris

Marius Wines (McLaren Vale)
www.mariuswines.com.au
Mourvedre

Maroochy Springs (Queensland Coastal)
www.sunshinecoastwine.com.au
Chambourcin, Verdelho

Marri Wood Park (Margaret River)
www.margaretriver.com
Chenin blanc

Marschall Groom Cellars (Barossa Valley)
www.groomwines.com
Zinfandel

Mary Byrnes Wines (Granite Belt)
www.marybryneswine.com
Grenache, Marsanne, Mourvedre, Roussanne, Viognier

Marybrook Vineyards & Winery (Margaret River)
Bastardo, Gamay, Grenache, Verdelho

mas serrat (Yarra Valley)
www.serrat.com.au
Viognier, Grenache

Mason Wines (Granite Belt)
www.masonwines.com.au
Petit verdot, Verdelho, Viognier

Massena Wines (Barossa Valley)
www.massena.com.au
Barbera, Cinsaut, Dolcetto, Durif, Grenache, Mourvedre, Roussanne, Saperavi, Tannat, Tinta amarela

Massoni (Pyrenees)
www.massoniwines.com
Barbera, Sangiovese

Maverick Wines (Barossa Valley)
www.maverickwines.com.au
Grenache, Mourvedre

Mawson Ridge (Adelaide Hills)
www.mawsonridge.com
Grenache, Pinot gris

Maxwell Wines (McLaren Vale)
www.maxwellwines.com.au
Grenache, Verdelho, Viognier

Mayfield Vineyard (Orange)
www.mayfieldvineyard.com
Sangiovese

Mayford Wines (Alpine Valleys)
www.mayfordwines.com
Tempranillo

Mazza (Geographe)
www.mazzawines.com
Bastardo, Graciano, Souzao, Tempranillo, Tinto Cao, Touriga

McAdams Lane (Geelong)
www.mcadamslane.com.au
Picolit, Pinot gris, Zinfandel

McCrae Mist Wines (Mornington Peninsula)
Pinot gris, Sangiovese

McCuskers Vineyard (Perth Hills)
Carnelian, Malbec, Tempranillo, Verdelho, Viognier

McHenry Hohnen (Margaret River)
www.mchv.com.au
Grenache, Marsanne, Malbec, Mourvedre, Petit verdot, Roussanne, Tempranillo, Viognier

McIvor Creek (Heathcote)
Marsanne, Viognier

McIvor Estate (Heathcote)
www.mcivorestate.com.au
Marsanne, Nebbiolo, Roussanne, Sangiovese

McKellar Ridge (Canberra)
www.mckellarridgewines.com.au
Viognier

McLaren Ridge Estate (McLaren Vale)
www.mclarenridge.com
Grenache

McLaren Wines (McLaren Vale)
www.mclarenwines.com
Grenache, Pinot gris

McLeish Estate (Hunter Valley)
www.mcleishhunterwines.com.au
Verdelho

McPherson Wines (Nagambie Lakes)
www.mcphersonwines.com.au
Marsanne, Mourvedre, Verdelho, Viognier

McVitty Grove (Southern Highlands)
www.mcvittygrove.com.au
Pinot gris

McWilliams (Riverina)
www.mcwilliams.com.au
Baco noir, Grenache, Tyrian, Verdelho

Meadowbank Estate (Southern Tasmania)
www.meadowbankwines.com.au
Meunier, Pinot gris

WINERIES • M

Meera Park (Hunter Valley)
www.meereapark.com.au
Verdelho, Viognier

Melaleuca Grove (Upper Goulburn)
www.melalleucawines.com.au
Marsanne

Melange Wines (Riverina)
Durif, Trebbiano, Verdelho

Melross Estate (Pyrenees)
www.melrossestate.com
Pinot gris, Sangiovese

Melville Hill Estate Wines (New England)
www.melvillehill.com.au
Tempranillo, Verdelho

Merum (Pemberton)
www.merum.com.au
Viognier

Metier Wines (Yarra Valley)
www.metierwines.com.au
Viognier

Miceli (Mornington Peninsula)
Pinot gris

Michael Unwin Wines (Grampians)
http;//www.michaelunwinwines.com.au
Barbera, Durif, Sangiovese

Michelini (Alpine Valleys)
www.micheliniwines.com.au
Barbera, Fragola, Marzemino, Pinot gris, Sangiovese, Teroldego

Middlebrook Estate (McLaren Vale)
www.middlebrookestate.com.au
Barbera, Chenin blanc, Grenache

Middlesex 31 (Great Southern)
Verdelho

Midhill Vineyard (Macedon Ranges)
www.macedonrangeswine.com.au
Gewurztraminer

Mihi Creek Vineyard (New England)
home.bluepin.net.au/mihicreek
Viognier

Miles from Nowhere (Margaret River)
www.milesfromnowhere.com.au
Petit verdot, Viognier

Millbrook Winery (Perth Hills)
www.millbrookwinery.com.au
Malbec, Petit verdot, Viognier

Milldale Estate Vineyard (Hunter Valley)
www.milldale.com.au
Viognier

Millers Dixons Creek Estate (Yarra Valley)
www.graememillerwines.com.au
Petit verdot, Pinot gris

Milton Vineyard (Southern Tasmania)
www.miltonvineyard.com.au
Gewurztraminer, Pinot gris

Minko (Southern Fleurieu)
www.minkowines.com
Pinot gris

Minnow Creek (McLaren Vale)
www.minnowcreekwines.com.au
Malbec, Sangiovese

Miramar (Mudgee)
www.miramarwines.com.au
Gewurztraminer

Miranda Wines, Griffith (Riverina)
www.mirandawines.com.au
Durif

Miranda (Riverina)
www.mirandawines.com.au
Durif

Mitchell (Clare Valley)
www.mitchellwines.com
Grenache, Mourvedre, Sangiovese

Mitchelton (Nagambie Lakes)
www.mitchelton.com.au
Grenache, Marsanne, Mourvedre, Roussanne, Viognier

WINERIES • M

Moama Wines (Perricoota)
members.iinet.net.au/~mcdonro
Chenin blanc, Petit verdot, Sangiovese, Verdelho

Moaning Frog (Margaret River)
www.moaningfrog.com.au
Viognier

Moffatdale Ridge (South Burnett)
www.moffatdaleridge.com.au
Verdelho

Molly's Cradle (Hunter Valley)
www.mollyscradle.com.au
Verdelho

Molly Morgan Vineyard (Hunter Valley)
www.mollymorgan.com
Verdelho

Mollydooker Wines (South Australia)
www.mollydookerwines.com.au
Verdelho

Monahan Estate (Hunter Valley)
www.monahanestate.com.au
Verdelho

Monichino Wines (Goulburn Valley)
www.monichino.com.au
Barbera, Orange muscat, Pinot gris, Sangiovese

Montalto Vineyards (Mornington Peninsula)
www.montalto.com.au
Meunier, Pinot gris

Montefalco Vineyard (Porongurup)
www.montefalco.com.au
Cabernet franc, Nebbiolo, Sangiovese

Monument Vineyard (Central Ranges Zone)
www.monumentvineyard.com.au
Barbera, Marsanne, Nebbiolo, Pinot gris, Ruby Cabernet, Sangiovese, Verdelho

Moondah Brook (Swan Valley)
www.moondahbrook.com.au
Chenin blanc, Verdelho

Moondarra (Gippsland)
Nebbiolo, Picolit, Pinot gris

Moorebank Vineyard (Hunter Valley)
www.moorebankvineyard.com.au
Gewurztraminer

Moorilla Estate (Southern Tasmania)
www.moorilla.com.au
Gewurztraminer, Pinot gris

Moorooduc Estate (Mornington Peninsula)
www.moorooduc-estate.com.au
Pinot gris

Mopoke Ridge Winery (Shoalhaven Coast)
www.mopokeridge.com.au
Chambourcin, Sangiovese, Verdelho

Moppa Wilton Vineyards (Barossa Valley)
Grenache

Moppity Vineyards (Hilltops)
www.moppity.com.au
Nebbiolo, Sangiovese, Viognier

MorganField (Macedon Ranges)
www.morganfield.com.au
Meunier

Morning Star Estate (Mornington Peninsula)
www.morningstarestate.com.au
Pinot gris

Morning Sun Vineyard (Mornington Peninsula)
www.morningsunvineyard.com.au
Barbera, Pinot gris

Morris (Rutherglen)
www.morriswines.com.au
Cinsaut, Durif, Muscadelle

Morrisons of Glenrowan (Glenrowan)
www.morrisonsofglenrowan.com
Durif, Tempranillo, Viognier, Zinfandel

WINERIES • M

Morrisons Riverview Winery (Perricoota)
www.riverviewestate.com.au
Grenache

Mosquito Hill Wines (Southern Fleurieu)
Pinot blanc, Savagnin

Moss Brothers (Margaret River)
www.mossbrothers.com.au
Graciano, Grenache, Tempranillo, Verdelho

Mount Anakie (Geelong)
Biancone, Lexia

Mount Appallan Vineyards (South Burnett)
www.mtappallan.com.au
Grenache, Mourvedre, Petit verdot, Verdelho, Viognier

Mount Ashby Estate (Southern Highlands)
www.mountashby.com.au
Pinot gris

Mount Avoca (Pyrenees)
www.mountavoca.com
Cabernet franc, Tempranillo, Viognier

Mount Broke Wines (Hunter Valley)
www.MtBrokeWines.com.au
Barbera, Verdelho

Mount Buffalo Vineyard (Alpine Valleys)
Viognier

Mount Burrumboot Estate (Heathcote)
www.burrumboot.com
Gamay, Marsanne, Petit verdot, Sangiovese, Tempranillo, Verdelho, Viognier

Mount Camel Ridge Estate (Heathcote)
www.mountcamelridgeestate.com
Mourvedre, Petit verdot, Viognier

Mount Charlie Winery (Macedon Ranges)
www.mountcharlie.com.au
Malbec, Tempranillo

Mount Cole Wineworks (Grampians)
www.mountcolewineworks.com.au
Viognier, Nebbiolo

Mount Eyre Vineyards (Hunter Valley)
www.mounteyre.com
Cabernet franc, Chambourcin

Mount Franklin Estate (Macedon Ranges)
www.mtfranklinwines.com.au
Dolcetto, Nebbiolo, Pinot gris

Mount Langi Ghiran Vineyards (Grampians)
www.langi.com.au
Barbera, Sangiovese, Pinot gris

Mount Macedon Winery (Macedon Ranges)
www.mountmacedonwinery.com.au
Gewurztraminer, Meunier

Mount Majura (Canberra)
www.mountmajura.com.au
Graciano, Pinot gris, Tempranillo

Mount Markey (Gippsland)
www.omeo.org.au
Pinot gris

Mount Pierrepoint Estate (Henty)
www.mountpierpoint.com
Pinot gris

Mount Pilot Estate (North East Victoria)
www.mtpilotestatewines.com.au
Durif, Viognier

Mount Prior (Rutherglen)
www.rutherglenvic.com
Durif, Muscadelle

Mount Surmon (Clare Valley)
www.mtsurmon.com.au
Nebbiolo, Pinot gris, Viognier

Mount Tamborine (Queensland Coastal)
www.tamborineestate.com.au
Verdelho

Mount Towrong (Macedon Ranges)
Nebbiolo, Prosecco

Mount Trafford (Southern Fleurieu)
Grenache

Mount Trio Vineyard (Porongurup)
www.mounttriowines.com.au
Viognier

Mount View Estate (Hunter Valley)
www.mtviewestate.com.au
Verdelho

Mount Vincent Estate (Hunter Valley)
www.mvewines.com.au
Sangiovese, Verdelho

Mountadam (Eden Valley)
www.mountadam.com.au
Pinot gris, Viognier

Mountilford (Mudgee)
Sylvaner

Mr Riggs Wine Company (McLaren Vale)
www.pennyshill.com.au
Petit verdot, Tempranillo, Viognier

Mt Billy (Southern Fleurieu)
www.mtbillywines.com.au
Grenache, Meunier, Mourvedre

Mt Samaria Vineyard (Goulburn Valley)
www.m-s-v.com.au
Pinot gris, Tempranillo

Mudgee Growers (Mudgee)
www.mudgeegrowers.com.au
Petit verdot, Verdelho

Mudgee Wines (Mudgee)
www.mudgeewines.com.au
Chambourcin, Durif, Gewurztraminer, Petit verdot, Trebbiano

Mulcra Estate Wines (Murray Darling)
www.mulcraestate.com.au
Petit verdot

Mulligan Wongara Vineyard (Cowra)
Sangiovese

Mulyan (Cowra)
www.mulyanwines.com.au
Sangiovese, Viognier

Munari (Heathcote)
www.munariwines.com
Malbec, Marsanne, Viognier

Mundoonen (Canberra)
www.mundoonen.com.au
Viognier

Mundrakoona Estate (Southern Highlands)
www.mundrakoona.com.au
Tempranillo

Murdock (Barossa Valley)
www.murdockwines.com
Grenache, Tempranillo

Murdup Wines (Mount Benson)
www.murdupwines.com.au
Pinot gris, Verdelho

Murray Estate (North East Victoria)
www.murrayestatewines.com.au
Chenin blanc

Murray Street Vineyard (Barossa Valley)
www.murraystreet.com.au
Cinsaut, Grenache, Marsanne, Mourvedre, Viognier, Zinfandel

Murrumbateman Winery (Canberra)
murrumbatemanwines.com.au
Verdelho

Myalup Wines (Geographe)
www.myalupvines.com
Chenin blanc

WINERIES • N

Myattsfield Vineyard and Winery (Perth Hills)
www.myattsfield.com.au
Durif, Mourvedre, Touriga, Verdelho, Viognier

Mylkappa Wines (Adelaide Hills)
www.mylkappa.com.au
Pinot gris

Myrtle Vale Vineyard (Upper Goulburn)
Viognier

N

Nalbra Estate (Geelong)
Pinot gris, Viognier

Nashwauk (McLaren Vale)
www.nashwaukvineyards.com.au
Tempranillo

Nazaaray (Mornington Peninsula)
www.nazaaray.com.au
Pinot gris

Neagles Rock Vineyards (Clare Valley)
www.neaglesrock.com
Grenache, Sangiovese

Neilson Estate Wines (Swan Valley)
www.neilsonestate.com.au
Verdelho

Nelwood Wines (Riverland)
www.nelwood.com
Petit verdot

Nepenthe (Adelaide Hills)
www.nepenthe.com.au
Pinot gris, Tempranillo, Viognier, Zinfandel

Neqtar Wines (Murray Darling)
www.neqtar.com.au
Pinot gris, Viognier

New Glory (Goulburn Valley)
Durif, Mourvedre, Petit verdot, Sangiovese, Verdelho

Nightingale Wines (Hunter Valley)
www.nightingalewines.com.au
Chambourcin, Verdelho

No Regrets (Southern Tasmania)
Gewurztraminer

Noon Winery (McLaren Vale)
www.noonwinery.com.au
Grenache

Noorilim Estate (Goulburn Valley)
www.noorilimestate.com.au
Viognier

Noorinbee Selection Vineyards (Gippsland)
Harslevelu, Malbec, Mammolo, Petit verdot

Noosa Valley Winery (Queensland Coastal)
Chambourcin

Normanby Wines (Queensland Zone)
www.normanbywines.com.au
Chambourcin, Durif, Grenache, Verdelho, Viognier

Norse Wines (Queensland Coastal)
www.norsewines.com
Colombard, Touriga, Verdelho

Nova Vita Wines (Adelaide Hills)
www.novavitawines.com.au
Pinot gris, Sangiovese

Nowra Hill Vineyard (Shoalhaven Coast)
Chambourcin, Malbec, Verdelho

Nugan Estate (King Valley)
www.nuganestate.com.au
Durif, Pinot gris, Sangiovese, Tempranillo

Nursery Ridge (Murray Darling)
Petit verdot, Viognier

Nyora Vineyard and Winery (Gippsland)
Colombard, Pinot gris, Verdelho

O

O'Donohue's Find (Riverland)
www.tomsdrop.com.au
Mourvedre

O'Regan Creek Vineyard and Winery (Queensland Coastal)
Chambourcin, Zinfandel

Oak Dale Wines (Swan Hill)
Mourvedre

Oak Works (Riverland)
Chambourcin, Durif, Montepulciano, Nebbiolo, Pinotage, Saperavi, Tannat, Tinta Negra Mole, Zinfandel

Oakover Estate (Swan Valley)
www.oakoverwines.com.au
Chenin blanc, Verdelho, Viognier

Oakridge (Yarra Valley)
www.oakridgeestate.com.au
Pinot gris

Oakvale (Hunter Valley)
www.oakvalewines.com.au
Chambourcin, Sangiovese, Verdelho

Oatley Wines (Mudgee)
www.oatleywines.com.au
Barbera, Pinot gris, Viognier

Ocean Eight Vineyard and Winery (Mornington Peninsula)
www.oceaneight.com
Pinot gris

Oceanview Estates (Queensland Coastal)
www.oceanviewestates.com.au
Ruby Cabernet, Viognier

Old Caves (Granite Belt)
www.oldcaveswinery.com.au
Zinfandel

Old Loddon Wines (Bendigo)
Cabernet franc

Old Mill Estate (Langhorne Creek)
www.outbacchus.com
Touriga

Old Plains (Adelaide Plains)
www.oldplains.com
Grenache

Old Station (Clare Valley)
Grenache

Olive Farm (Swan District)
www.olivefarm.com.au
Chenin blanc, Grenache, Verdelho

Oliverhill (McLaren Vale)
Grenache

Olivers Taranga (McLaren Vale)
www.oliverstaranga.com
Fiano, Grenache, Sagrantino, Tempranillo, Viognier

Olsen (Margaret River)
www.olsen.com.au
Verdelho

Olssens of Watervale (Clare Valley)
Carmenere, Malbec, Mourvedre, Petit verdot, Zinfandel

Omersown Wines (Riverland)
www.omersownwines.com.au
Savagnin

Optimiste (Mudgee)
www.optimiste.com.au
Petit verdot, Pinot gris, Tempranillo, Viognier

Orange Mountain (Orange)
www.orangemountain.com.au
Viognier

Orchard Road (Orange)
Barbera, Pinot gris

Organic Vignerons Australia (Riverland)
www.ova.com.au
Grenache, Mourvedre, Viognier

Orlando (Barossa Valley)
www.orlandowines.com
Grenache, Rubienne

Outlook Hill (Yarra Valley)
www.outlookhill.com.au
Pinot gris

WINERIES • P

Outram Estate (Hunter Valley)
Verdelho

P

Palmara (Southern Tasmania)
www.palmara.com.au
Siegerrebe

Palmers Wines (Hunter Valley)
www.palmerwines.net.au
Verdelho

Pangallo Estate (Hunter Valley)
www.pangalloestate.com.au
Zibibbo

Paracombe Wines (Adelaide Hills)
www.paracombewines.com
Cabernet franc, Pinot gris, Viognier

Paradigm Hill (Mornington Peninsula)
www.paradigmhill.com.au
Pinot gris

Paradine Estate (Queensland Zone)
Durif, Mourvedre

Paramoor Wines (Macedon Ranges)
www.paramoor.net.au
Pinot gris

Paringa Estate (Mornington Peninsula)
www.paringaestate.com.au
Pinot gris

Parish Hill Wines (Adelaide Hills)
www.parishhillwines.com.au
Arneis, Dolcetto, Fiano, Nebbiolo, Negro amaro, Vermentino

Parkerville Ponds Vineyard (Perth Hills)
Grenache

Parri Estate (Southern Fleurieu)
www.parriestate.com.au
Grenache, Viognier

Passing Clouds (Bendigo)
www.totinowines.com.au
Cabernet franc

Pasut Family Wines (Murray Darling)
Barbera, Fragola, Nebbiolo, Pinot gris, Sangiovese, Vermentino

Patrice Winery (North East Victoria)
www.patricewines.com.au
Durif, Petit verdot

Paul Bettio (King Valley)
www.paulbettiowines.com.au
Barbera, Pinot gris

Paul Conti Wines (Greater Perth Zone)
www.paulcontiwines.com.au
Chenin blanc, Grenache

Paulmara Estate (Barossa Valley)
www.paulmara.com.au
Sangiovese

Paxton (McLaren Vale)
www.paxtonvineyards.com
Grenache, Pinot gris, Tempranillo

Peel Estate (Peel)
www.peelwine.com.au
Chenin blanc, Souzao, Tinto Cao, Tinta amarela, Touriga, Verdelho, Zinfandel

Peel Ridge (Peel)
www.peelridge.com.au
Nebbiolo, Tempranillo, Verdelho, Viognier

Peerick Vineyard (Pyrenees)
www.peerick.com.au
Viognier

Pegeric (Macedon Ranges)
www.pegeric.com
Barbera

Pelican's Landing Maritime Wines (Southern Fleurieu)
www.plmwines.com
Viognier

Pende Valde (McLaren Vale)
www.pendevalde.com.au
Grenache

Peninsula Baie Wines (Geelong)
Pinot gris

Penmara (Hunter Valley)
www.penmarawines.com.au
Petit verdot, Sangiovese, Verdelho

Pennan (Margaret River)
www.pennanwines.com.au
Viognier

Penny's Hill (McLaren Vale)
www.pennyshill.com.au
Grenache, Marsanne, Roussanne, Verdelho, Viognier

Pennyfield Wines (Riverland)
www.pennyfieldwines.com.au
Mourvedre, Petit verdot, Touriga, Viognier

Pennyweight Winery (Beechworth)
www.pennyweight.com.au
Gamay, Palomino, Tinto Cao, Touriga

Peos Estate (Manjimup)
www.peosestate.com.au
Carnelian, Verdelho

Pepper Tree Wines (Orange)
www.peppertreewines.com.au
Aglianico, Cabernet franc, Gewurztraminer, Tannat, Verduzzo, Viognier, Zinfandel

Pepperilly Estate Wines (Geographe)
www.pepperilly.com
Grenache, Viognier

Peppin Ridge (Upper Goulburn)
Marsanne, Verdelho

Perrini Estate (Adelaide Hills)
www.perriniwines.com.au
Sangiovese

Pertaringa (McLaren Vale)
www.pertaringa.com.au
Aglianico, Grenache, Moscato, Nero d'Avola, Petit verdot, Tannat, Verdelho

Petaluma (Adelaide Hills)
www.petaluma.com.au
Pinot gris, Viognier

Peter Lehmann (Barossa Valley)
www.peterlehmanwines.com
Cabernet franc, Chenin blanc, Grenache, Malbec, Mourvedre, Muscadelle, Tempranillo

Petersons Glenesk Estate (Mudgee)
www.petersonswines.com.au
Chambourcin, Durif, Gewurztraminer, Petit verdot, Verdelho, Viognier, Zinfandel

Pettavel (Geelong)
www.pettavel.com
Petit verdot, Viognier

Pewsy Vale (Eden Valley)
www.pewseyvale.com
Gewurztraminer, Pinot gris

Pfeiffer Wines (Rutherglen)
www.pfeifferwines.com.au
Gamay, Marsanne, Muscadelle

Phaedrus Estate (Mornington Peninsula)
www.phaedrus.com.au
Pinot gris

Philip Lobley Wines (Upper Goulburn)
Nebbiolo

Philip Shaw (Orange)
www.philipshaw.com.au
Viognier

Phillips Estate (Pemberton)
Zinfandel

Phoenix Estate (Clare Valley)
Grenache, Mourvedre, Pedro Ximenez, Petit verdot, Viognier

Piako Vineyards (Murray Darling)
www.chalmersnurseries.com.au
Durif, Lagrein, Petit verdot, Pinot gris, Sangiovese, Tempranillo

WINERIES • P 175

Pialligo Estate (Canberra)
www.pialligoestate.com.au
Pinot gris, Sangiovese

Piano Gully (Manjimup)
www.pianogully.com.au
Viognier

Pier 10 (Mornington Peninsula)
www.pier10.com.au
Pinot gris, Ruby Cabernet

Pieter van Gent (Mudgee)
www.pvgwinery.com.au
Muller Thurgau, Verdelho

Piggs Peake Winery (Hunter Valley)
www.piggspeake.com
Barbera, Marsanne, Sangiovese, Verdelho, Zinfandel

Pike and Joyce (Adelaide Hills)
www.pikeswines.com.au
Pinot gris

Pikes (Clare Valley)
www.pikeswines.com.au
Chenin blanc, Grenache, Mourvedre, Pinot gris, Sangiovese, Tempranillo, Viognier

Pindarie Wines (Barossa Valley)
www.pindarie.com.au
Sangiovese, Tempranillo

Pinelli (Swan Valley)
Chenin blanc, Verdelho

Pipers Brook Vineyard (Northern Tasmania)
www.pbv.com.au
Gewurztraminer, Pinot gris

Pirie Estate (Northern Tasmania)
www.pirietasmania.com.au
Gewurztraminer, Pinot gris

Piromit Wines (Riverina)
www.piromitwines.com.au
Colombard, Durif, Pinot gris, Sangiovese, Verdelho

Pirramimma (McLaren Vale)
www.pirramimma.com.au
Grenache, Petit verdot, Tannat, Viognier

Pizzini Wines (King Valley)
www.pizzini.com.au
Arneis, Brachetto, Nebbiolo, Picolit, Sangiovese, Verduzzo

Plan B (Margaret River)
www.planbwines.com
Tempranillo, Viognier

Plantagenet (Mount Barker)
www.plantagenetwines.com
Cabernet franc, Chenin blanc, Grenache

Platt's (Rutherglen)
Gewurztraminer

Plunkett Fowles (Strathbogie Ranges)
www.plunkettfowles.com.au
Gewurztraminer, Savagnin, Viognier

Poachers Ridge Vineyards (Mount Barker)
www.prv.com.au
Marsanne, Viognier

Poet's Corner (Mudgee)
www.poetscornerwines.com.au
Barbera, Sangiovese

Point Leo Road Vineyard (Mornington Peninsula)
www.pointleoroad.com.au
Gewurztraminer, Lagrein, Pinot gris

Pokolbin Estate (Hunter Valley)
www.pokolbinestate.com.au
Nebbiolo, Sangiovese, Tempranillo, Verdelho

Polin & Polin (Hunter Valley)
www.polinwines.com.au
Verdelho

Politini (King Valley)
www.politiniwines.com.au
Grecanico, Pinot gris, Sangiovese, Vermentino

176 WINERIES • Q

Polleters Vineyard (Pyrenees)
www.polleters.com
Cabernet franc

Pollocksford Vineyards (Geelong)
Chenin blanc

Pondalowie (Bendigo)
www.pondalowie.com.au
Malbec, Tempranillo, Touriga, Viognier

Pooley Wines (Southern Tasmania)
www.pooleywines.com.au
Pinot gris

Port Phillip Estate (Mornington Peninsula)
www.portphillip.net
Arneis, Barbera

Port Robe (Limestone Coast Zone)
Cygne blanc

Portree Vineyard (Macedon Ranges)
www.portreevineyard.com.au
Cabernet franc

Possums Vineyard (McLaren Vale)
www.possumswines.com.au
Grenache, Viognier

Pothana (Hunter Valley)
www.davidhookwines.com.au
Pinot gris, Verdelho, Viognier

Prancing Horse Estate (Mornington Peninsula)
www.prancinghorseestate.com
Pinot gris

Preston Peak (Granite Belt)
www.prestonpeak.com
Verdelho

Primerano (King Valley)
www.primerano.com.au
Pinot gris

Primo Estate (Adelaide Plains)
www.primoestate.com.au
Barbera, Colombard, Nebbiolo, Pinot gris, Sangiovese

Prince Hill Wines (Mudgee)
www.princehillwines.com
Barbera, Pinot gris, Sangiovese, Verdelho

Prince of Orange (Orange)
www.princeoforangewines.com.au
Viognier

Printhie Wines (Orange)
www.printhiewines.com.au
Pinot gris, Viognier

Protero (Adelaide Hills)
www.proterowines.com.au
Nebbiolo, Viognier

Provenance Wines (Geelong)
www.provenancewines.com.au
Pinot gris

Punt Road (Yarra Valley)
www.puntroadwines.com.au
Pinot gris

Purple Hen Wines (Gippsland)
www.purplehenwines.com.au
Viognier

Pycnantha Hill Estate (Clare Valley)
www.pycnanthahill.com.au
Sangiovese

Pyramid Gold (Bendigo)
Mourvedre

Pyramid Hill Wines (Hunter Valley)
www.pyramidhillwines.com
Verdelho

Pyramids Road Wines (Granite Belt)
www.pyramidsroad.com.au
Mourvedre, Verdelho

Pyren Vineyard (Pyrenees)
www.pyrenvineyard.com
Durif, Malbec, Petit verdot, Viognier

Q

Quarry Hill Wines (Canberra)
www.quarryhill.com.au
Savagnin, Tempranillo

WINERIES • R

Quattro Mano (Barossa Valley)
www.quattromano.com.au
Tempranillo, Touriga

Quealy (Mornington Peninsula)
www.quealy.com.au
Pinot gris, Sangiovese, Tocai fruilano

Quoin Hill (Pyrenees)
www.quoinhill.com.au
Pinot gris, Tempranillo

R

Racecourse Lane Wines (Hunter Valley)
www.racecourselane.com.au
Sangiovese, Verdelho, Viognier

Rahona Valley Vineyard (Mornington Peninsula)
www.rahonavalley.com.au
Meunier

Raleigh Wines (Northern Rivers Zone)
www.raleighwines.com
Chambourcin, Villard blanc

Ralph Fowler Wines (Mount Benson)
www.ralphfowlerwines.com.au
Viognier

Rangemore Estate (Darling Downs)
rangemoreestate.com.au
Verdelho, Viognier

Ravens Croft Wines (Granite Belt)
www.ravenscroftwines.com.au
Petit verdot, Verdelho

Ravensworth Wines (Canberra)
www.ravensworthwines.com.au
Marsanne, Sangiovese, Viognier

RBJ (Barossa Valley)
www.rjbwines.com
Grenache, Mourvedre

Reads (King Valley)
Crouchen

Red Cliffs (Murray Darling)
Petit verdot

Red Earth Estate (Western Plains)
www.redearthestate.com.au
Barbera, Carmenere, Gamay, Grenache, Tempranillo, Torrentes, Verdelho

Red Edge (Heathcote)
Tempranillo, Monastrell

Red Hill Estate (Mornington Peninsula)
www.redhillestate.com.au
Meunier, Pinot gris

Red Mud (Riverina)
www.nelwood.com
Petit verdot

Red Tail (Northern Rivers Zone)
Colombard, Verdelho

Redbank Victoria (King Valley)
www.redbankwines.com
Moscato, Pinot gris, Viognier

Redbox (Yarra Valley) (Yarra Valley)
www.redboxvineyard.com.au
Pinot gris

Redbox Perricoota (Perricoota)
www.redboxvineyard.com.au
Barbera, Grenache, Mourvedre

Redgate (Margaret River)
www.redgatewines.com.au
Cabernet franc, Chenin blanc, Verdelho

Redheads Studio (McLaren Vale)
www.redheadswine.com.au
Grenache, Viognier

Reedy Creek (Northern Slopes Zone)
www.reedycreekwines.com.au
Durif, Malbec, Mourvedre

Rees Miller Estate (Upper Goulburn)
www.reesmiller.com
Viognier

Reg Drayton (Hunter Valley)
www.regdraytonwines.com.au
Verdelho

Regent Wines (Swan Hill)
Verdelho, Viognier

Reilly's Wines (Clare Valley)
www.reillyswines.com
Grenache

Remarkable View Winery (Southern Flinders Region)
www.remarkableview.com.au
Grenache, Petit verdot, Sangiovese, Tempranillo

Renewan (Swan Hill)
www.murrayriverwines.com.au
Durif

Richard Hamilton Wines (McLaren Vale)
www.hamiltonwines.com
Grenache

Richfield Estate (New England)
www.richfieldvineyard.com.au
Ruby Cabernet, Verdelho

Ridgemill Estate (Granite Belt)
www.ridgemillestate.com
Lagrein, Saperavi, Tempranillo, Verdelho, Viognier

Ridgeview Wines (Hunter Valley)
www.ridgeview.com.au
Chambourcin, Gewurztraminer, Pinot gris, Verdelho, Viognier

Rigel Wines (Mornington Peninsula)
www.rigelwines.com.au
Nebbiolo, Sangiovese, Zinfandel

Rimfire Vineyards (Darling Downs)
www.rimfirewinery.com.au
1893, Aglianico, Aleatico, Cabernet franc, Colombard, Graciano, Malvasia, Marsanne, Muscat, Ruby Cabernet, Sangiovese, Touriga, Verdelho, Vermentino

Ringer Reef Winery (Alpine Valleys)
www.ringerreef.com.au
Sangiovese

Riseborough Estate (Swan District)
www.riseborough.com.au
Chenin blanc, Grenache

Rivendell (Margaret River)
www.rivendellwines.com.au
Souzao, Tinto Cao, Touriga, Verdelho

Riverbank Estate (Swan Valley)
www.riverbankestate.com.au
Chenin blanc, Grenache, Malbec, Petit verdot, Sangiovese, Tempranillo, Verdelho, Viognier, Zinfandel

Riverina Estate Wines (Riverina)
www.riverinaestate.com
Barbera, Chenin blanc, Durif, Marsanne, Verdelho

Riversands Winery (Queensland Zone)
www.riversandswines.com
Aleatico, Jacquez, Ruby Cabernet

Robert Channon (Granite Belt)
www.robertchannonwines.com
Verdelho

Robert Johnson Vineyards (Eden Valley)
robertjohnsonvineyards.com.au
Viognier

Robert Stein (Mudgee)
www.robertstein.com
Gewurztraminer

Roberts Estate (Murray Darling)
www.robertsestatewines.com
Petit verdot, Sangiovese, Verdelho, Viognier

Robertson of Clare (Clare Valley)
www.rocwines.com.au
Malbec, Petit verdot

Robinvale Wines (Murray Darling)
www.organicwines.com
Kerner, Lexia, Mavrodaphne, Ruby Cabernet, Trollinger, Zinfandel

WINERIES • R

Robyn Drayton (Hunter Valley)
www.robyndrayton.com.au
Petit verdot, Verdelho

Rochford Wines (Yarra Valley)
www.rochfordwines.com.au
Arneis, Pinot gris

Rockford (Barossa Valley)
www.rockfordwines.com.au
Alicante Bouchet, Grenache, Mourvedre

Rocky Passes Wines (Upper Goulburn)
www.rockypasseswines.com.au
Viognier

Rodericks (South Burnett)
Colombard, Malbec, Tarrango, Verdelho

Roehr (Barossa Valley)
Grenache

Roennfeldt Wines (Barossa Valley)
Grenache

Rogues Lane Vineyard (Heathcote)
www.rougueslane.com.au
Malbec

Rojo Wines (Port Phillip Zone)
Dolcetto, Nebbiolo, Sangiovese

Romantic Vineyard (Pyrenees)
www.romanticvineyard.com
Petit verdot, Tempranillo

Romavilla (Roma)
www.romavilla.com
Chenin blanc, Crouchen, Garganega, Ruby Cabernet, Viognier, Zinfandel

Rookery Wines (Kangaroo Island)
www.rookerywines.com.au
Petit verdot, Sangiovese, Saperavi, Tempranillo

Rose Hill Estate Wines (King Valley)
rosehillestatewines.com.au
Durif

Rosebrook Estate (Hunter Valley)
www.hunterriverretreat.com.au
Verdelho

Roselea Estate (Shoalhaven Coast)
www.roseleavineyard.com.au
Chambourcin, Nebbiolo, Verdelho

Rosenvale Wines (Barossa Valley)
www.rosenvale.com.au
Grenache

Rosevears Estate (Northern Tasmania)
www.rosevearsestate.com.au
Gewurztraminer, Pinot gris

Roslyn Estate (Southern Tasmania)
www.roslynestate.com.au
Petit verdot

Ross Estate Wines (Barossa Valley)
www.rossestate.com.au
Graciano, Grenache, Tempranillo

Ross Hill Wines (Orange)
www.rosshillwines.com.au
Cabernet franc

Rossiters (Murray Darling)
Barbera, Colombard, Lagrein, Vermentino

Rothbury Ridge (Hunter Valley)
www.rothburyridgewines.com.au
Chambourcin, Durif, Verdelho

Roundstone Winery (Yarra Valley)
www.yarravalleywine.com
Gamay, Viognier

Rowans Lane Wines (Henty)
www.rowanslanewines.com.au
Pinot gris

Rowanston on the Track (Macedon Ranges)
www.rowanston.com
Malbec, Viognier

Ruane Winery (Southern Highlands)
Pinot gris, Sangiovese

Rudderless Wines (McLaren Vale)
www.rudderlesswines.com.au
Graciano, Grenache, Malbec, Mourvedre, Viognier

Rumbarella (Granite Belt)
www.rumbalarawines.com.au
Verdelho

Rupert's Ridge Estate (Heathcote)
www.rupertsridge.com
Sagrantino, Vermentino, Viognier

Rusden Wines (Barossa Valley)
www.rusdenwines.com.au
Chenin blanc, Grenache, Mourvedre, Zinfandel

Rusticana (Langhorne Creek)
www.rusticanawines.com.au
Durif, Zinfandel

Rusty Fig Wines (South Coast Zone)
www.rustyfigwines.com.au
Chambourcin, Savagnin, Tempranillo, Verdelho

Rutherglen Estates (Rutherglen)
www.rutherglenestates.com.au
Arneis, Durif, Fiano, Grenache, Marsanne, Mourvedre, Nebbiolo, Sangiovese, Viognier, Zinfandel

Rymill Coonawarra (Coonawarra)
www.rymill.com.au
Gewurztraminer

S

Saddlers Creek Wines (Hunter Valley)
www.saddlerscreekwines.com.au
Verdelho

Sailors Falls Winery (Macedon Ranges)
www.sailorsfallsestate.com.au
Gamay, Gewurztraminer, Pinot gris

Saint Derycke's Wood (Southern Highlands)
www.saintderyckeswood.com.au
Marsanne

Salena Estate (Riverland)
www.salenaestate.com.au
Petit verdot

Salisbury Winery (Murray Darling)
Colombard, Petit verdot, Sangiovese, Viognier

Salitage (Pemberton)
www.salitage.com.au
Verdelho

Salomon Estate (Currency Creek)
www.salomonwines.com
Petit verdot

Saltram (Barossa Valley)
www.saltramwines.com.au
Grenache, Mourvedre, Viognier

Sam Miranda Wines (King Valley)
www.sammiranda.com.au
Durif, Moscato, Petit verdot, Prosecco, Verdelho

Samson Hill Estate (Yarra Valley)
Verdelho

Samuels Gorge (McLaren Vale)
www.gorge.com.au
Grenache, Tempranillo

Sand Hills Vineyard (Central Ranges Zone)
Colombard

Sandalford Wines (Swan Valley)
www.sandalford.com
Chenin blanc, Verdelho

Sandalyn Wilderness Estate (Hunter Valley)
www.huntervalleyboutiques.com.au
Verdelho

Sandhurst Ridge (Bendigo)
www.sandhurstridge.com.au
Nebbiolo

Sanguine Estate (Heathcote)
www.sanguine-estate.com.au
Petit verdot, Tempranillo, Viognier, Zinfandel

Sarabah Estate (Queensland Zone)
www.sarabahwines.com.au
Colombard, Verdelho, Viognier

Sarsfield Estate (Gippsland)
www.sarsfieldestate.com.au
Mourvedre

Sautjan Vineyards (Macedon Ranges)
sautjan.com
Pinot gris, Viognier

SC Pannell (McLaren Vale)
www.scpannell.com.au
Grenache, Nebbiolo, Pinot gris, Touriga

Scaffidi Estate (Adelaide Hills)
www.talunga.com.au
Nebbiolo, Sangiovese

Scarpatoni Estate (McLaren Vale)
www.scarpantoni-wines.com.au
Chenin blanc, Gamay, Grenache

Schild Estate Wines (Barossa Valley)
www.schildestate.com.au
Grenache, Mourvedre

Schiller Vineyards (Barossa Valley)
www.schillervineyards.com.au
Grenache

Schulz Vignerons (Barossa Valley)
Grenache, Mourvedre, Zinfandel

Schwarz Wine Company (Barossa Valley)
www.schwarzwineco.com.au
Grenache

Scion Vineyard (Rutherglen)
www.scionvineyard.com
Durif, Grenache, Orange muscat, Viognier

Scorpiiion (Barossa Valley)
www.scorpiiionwines.com.au
Grenache, Mourvedre

Scorpo Wines (Mornington Peninsula)
www.scorpowines.com.au
Pinot gris

Scotchmans Hill (Geelong)
www.scotchmanshill.com.au
Pinot gris

Seabrook Wines (Barossa Valley)
Viognier

Seaforth Vineyard (Mornington Peninsula)
www.seaforthwines.com.au
Pinot gris

Sedona Estate (Upper Goulburn)
www.sedonaestate.com.au
Sangiovese

Seldom Seen (Mudgee)
Gewurztraminer

Seppelt Great Western (Grampians)
www.seppelt.com.au
Marsanne, Ondenc, Palomino, Pinot gris, Roussanne

Serafino Wines (McLaren Vale)
www.serafinowines.com.au
Grenache, Nebbiolo, Tempranillo

Seraph's Crossing (Clare Valley)
Grenache, Mourvedre, Zinfandel

Serenella Estate (Hunter Valley)
www.serenella.com.au
Sangiovese, Verdelho

Settlement Wines (McLaren Vale)
www.settlementwines.com
Arneis, Cabernet franc, Palomino, Pedro Ximenez, Pinot gris, Tinta Negra Mole, Verdelho

Settlers Ridge (Margaret River)
www.settlersridge.com.au
Chenin blanc, Malbec, Sangiovese

Settlers Rise Montville (Queensland Coastal)
www.settlersrise.com.au
Verdelho

Seven Mile Vineyard (Shoalhaven Coast)
www.sevenmilevineyard.com.au
Chambourcin, Petit verdot, Verdelho

Seven Ochres (Margaret River)
www.sevenochres.com.au
Petit verdot, Viognier

Sevenhill Wines (Clare Valley)
www.sevenhillcellars.com.au
Barbera, Chenin blanc, Gewurztraminer, Grenache, Malbec, Muscadelle, Pedro Ximenez, Ruby Cabernet, Touriga, Verdelho, Viognier

Sevenoaks Wines (Hunter Valley)
www.sevenoakswines.com
Petit verdot, Sangiovese

Severn Brae Estate (Granite Belt)
www.severnbraewines.com
Sangiovese, Verdelho

Seville Estate (Yarra Valley)
www.sevilleestate.com.au
Pinot gris

Seville Hill (Yarra Valley)
www.sevillehill.com.au
Tempranillo

Shadowfax Vineyard and Winery (Geelong)
www.shadowfax.com.au
Pinot gris, Sangiovese, Viognier

Sharpe Wines of Orange (Orange)
www.sharpewinesoforange.com.au
Cabernet franc

Shawwood Estate (Mudgee)
www.shawwood.com.au
Verdelho

Shays Flat Vineyard (Pyrenees)
www.shaysflat.com
Sangiovese

Sheer Drop (Bendigo)
www.sheerdropwines.com.au
Pinot gris

Shelmerdine (Heathcote)
www.shelmerdinevineyards.com.au
Viognier

Shepherds Run (Canberra)
www.shepherdsrun.com.au
Gewurztraminer

Sherwood Estate (Hastings River)
www.sherwoodestatewines.com.au
Chambourcin, Sangiovese, Verdelho

Shingleback (McLaren Vale)
www.shingleback.com.au
Grenache

Sidewood Estate (Adelaide Hills)
www.sidewoodestate.com.au
Pinot gris

Sieber Road Wines (Barossa Valley)
www.sieberwines.com
Grenache, Mourvedre, Viognier

Sigismondi Estate Wines (Riverland)
www.southernsecret.com.au
Petit verdot

Silver Wings Winemaking (Goulburn Valley)
www.silverwingswines.com
Mourvedre

Silverfox Wines (Perricoota)
www.silverfoxwines.com.au
Mourvedre, Sangiovese, Verdelho

Silverwaters Vineyard (Gippsland)
Pinot gris

Simon Hackett (McLaren Vale)
Grenache

Sirromet (Queensland Coastal)
www.sirromet.com
Chambourcin, Marsanne, Mourvedre, Nebbiolo, Petit verdot, Pinot gris, Verdelho, Viognier

Sittella (Swan Valley)
www.sittella.com.au
Chenin blanc, Verdelho

Skillogalee (Clare Valley)
www.skillogalee.com.au
Gewurztraminer, Grenache, Malbec

Skimstone (Mudgee)
www.skimstone.com.au
Barbera, Sangiovese

Smallfry Wines (Barossa Valley)
www.smallfrywines.com.au
Carignan, Cinsaut, Grenache, Mourvedre, Muscadelle, Petit verdot, Tempranillo, Tinta amarela, Viognier

Smallwater Estate (Geographe)
www.smallwaterestate.com.au
Zinfandel

Smidge Wines (Langhorne Creek)
www.smidgewines.com
Viognier, Zinfandel

Smithbrook (Pemberton)
www.smithbrook.com.au
Petit verdot

SmithLeigh Vineyard (Hunter Valley)
Verdelho

Snobs Creek Wines (Upper Goulburn)
www.snobscreekvineyard.com.au
Dolcetto, Pinot gris, Roussanne, Viognier

Snowy Vineyard (Southern New South Wales Zone)
www.snowyvineyard.com
Muller Thurgau, Siegerrebe, Sylvaner

Solstice Mount Torrens Vineyards (Adelaide Hills)
www.solstice.com.au
Viognier

Somerbury Estate (Mornington Peninsula)
Pinot gris

Sons of Eden (Barossa Valley)
www.sonsofeden.com
Grenache, Mourvedre, Viognier

Sorby Adams (Eden Valley)
www.sorbyadamswines.com
Gewurztraminer, Viognier

Sorrenberg (Beechworth)
www.sorrenberg.com
Gamay

Soul Growers (Barossa Valley)
soulgrowers.com
Grenache, Mourvedre

Souters Vineyard (Alpine Valleys)
www.happyvalley75.com.au/soutersvineyard
Gewurztraminer

South Channel Wines (Mornington Peninsula)
Pinot gris

Southern Grand Estate (Hunter Valley)
Gewurztraminer, Verdelho

Southern Highland Wines (Southern Highlands)
www.southernhighlandwines.com
Chambourcin, Gewurztraminer, Moscato, Nebbiolo, Pinot gris, Sangiovese, Viognier

Spence Wines (Geelong)
www.spencewines.com.au
Viognier

Spinifex (Barossa Valley)
www.spinifexwines.com
Carignan, Cinsaut, Grenache, Grenache gris, Marsanne, Mourvedre, Trebbiano, Vermentino, Viognier

Splitrock Vineyard Estate (Hunter Valley)
www.splitrockvineyard.com.au
Verdelho

Spoehr Creek Wines (Adelaide Hills)
Viognier

Spook Hill Wines (Riverland)
www.spookhillwines.com
Grenache, Mourvedre

Spring Vale Wines (Southern Tasmania)
www.springvalewines.com
Gewurztraminer, Pinot gris

Springbrook Mountain Vineyard (Queensland Coastal)
www.springbrookvineyard.com
Verdelho

Springs Hill Vineyard (Fleurieu Zone)
www.springshill.com.au
Grenache, Mourvedre

St Annes Vineyards (Perricoota)
Grenache, Mourvedre

St Hallett (Barossa Valley)
www.sthallett.com.au
Grenache, Mourvedre, Touriga

St Huberts (Yarra Valley)
www.sthuberts.com.au
Roussanne

St Ignatius Vineyard (Pyrenees)
www.stignatiusvineyard.com.au
Sangiovese

St Leonards (Rutherglen)
www.stleonardswine.com.au
Cabernet franc, Chenin blanc, Orange muscat, Viognier

St Mary's (Limestone Coast Zone)
www.stmaryswines.com
Petit verdot

St Matthias (Northern Tasmania)
www.moorilla.com.au
Pinot blanc, Pinot gris

St Michael's Vineyard (Heathcote)
Petit verdot

St Petrox (Hunter Valley)
www.saintpetrox.com.au
Chambourcin, Durif, Mondeuse

Stakehill Estate (Peel)
Chenin blanc, Grenache, Taminga, Tarrango

Stanley Brothers (Barossa Valley)
www.stanleybros.mtx.net
Sylvaner

Stanton and Killeen Wines (Rutherglen)
www.stantonandkilleenwines.com.au
Durif, Muscadelle, Tempranillo, Tinto Cao, Touriga, Viognier

Stanton Estate (Queensland Zone)
Marsanne, Verdelho

Starvedog Lane (Adelaide Hills)
www.starvedoglane.com.au
Meunier, Nebbiolo, Pinot gris, Sangiovese, Tempranillo, Touriga

Staunton Vale Vineyard (Geelong)
Petit verdot

Steels Creek Estate (Yarra Valley)
www.steelsckestate.com.au
Cabernet franc, Colombard

Stefani Estate (Yarra Valley)
www.stefaniestatewines.com.au
Pinot gris

Stefano Lubiano (Southern Tasmania)
www.slw.com.au
Nebbiolo, Pinot gris

Steinborner Family Vineyards (Barossa Valley)
www.sfvineyards.com.au
Durif, Marsanne, Viognier

Stella Bella (Margaret River)
www.stellabella.com.au
Sangiovese, Tempranillo, Viognier

Stellar Ridge (Margaret River)
www.stellar-ridge.com
Verdelho, Zinfandel

Stevens Brook Estate (Perricoota)
www.stevensbrookwines.com
Petit verdot, Sangiovese, Verdelho

Sticks (Yarra Valley)
www.sticks.com.au
Viognier

WINERIES • S

Stirling Wines (Hunter Valley)
Verdelho

Stockman's Ridge (Central Ranges Zone)
www.stockmansridge.com.au
Pinot gris, Savagnin, Tempranillo

Stomp (Hunter Valley)
www.stompwine.com.au
Verdelho

Stone Bridge Estate (Manjimup)
Sangiovese

Stone Bridge Wines (Mount Lofty Ranges Zone)
Malbec, Pinot gris

Stone Coast Wines (Wrattonbully)
www.stonecoastwines.com
Pinot gris

Stone Ridge (Granite Belt)
www.stoneridgewine.com
Marsanne, Viognier

Stonehaven (Padthaway)
www.stonehavenvineyards.com.au
Gewurztraminer, Ruby Cabernet, Sangiovese, Viognier

Stonehurst Cedar Creek (Hunter Valley)
www.cedarcreekcottages.com.au
Chambourcin

Stonewell Vineyards (Barossa Valley)
wwww.stonewell.com.au
Grenache

Stringybark (Perth Hills)
Verdelho

Stroud Valley Wines (Northern Rivers Zone)
stroudvalleywines.com.au
Chambourcin, Verdelho

Stuart Range Estate (South Burnett)
www.stuartrange.com.au
Colombard, Verdelho

Stuart Wines (Heathcote)
www.stuartwinesco.com.au
Nebbiolo, Tempranillo, Viognier

Stumpy Gully (Mornington Peninsula)
www.stumpygully.com.au
Marsanne, Pinot gris, Sangiovese

Sugarloaf Ridge (Southern Tasmania)
www.sugarloafridge.com
Lagrein, Pinot gris, Viognier

Summerfield (Pyrenees)
www.summerfieldwines.com
Trebbiano

Summit Estate (Granite Belt)
www.summitestate.com.au
Malbec, Marsanne, Petit verdot, Tempranillo, Verdelho, Zinfandel

Surveyor's Hill Winery (Canberra)
www.survhill.com.au
Touriga

Sussanah Brook Wines (Swan District)
Chenin blanc, Malbec, Verdelho

Sutherland Estate (Yarra Valley)
www.sutherlandestate.com.au
Gewurztraminer, Tempranillo

Sutherlands Creek Vineyard (Geelong)
www.sutherlandscreek.com
Gamay, Grenache, Mourvedre, Pinot gris, Viognier, Zinfandel

Sutton Grange Winery (Bendigo)
www.suttongrangewines.com
Aglianico, Fiano, Sangiovese, Viognier

Swan Valley Wines (Swan Valley)
www.swanvalleywines.com.au
Chenin blanc, Grenache

Swanbrook Estate Wines (Swan Valley)
Chenin blanc, Verdelho

Sweet Water Hill Wines (Sunshine Coast)
Colombard

Swings & Roundabouts (Margaret River)
www.swings.com.au
Chenin blanc, Graciano, Grenache, Nebbiolo, Sangiovese, Tempranillo, Viognier

Swooping Magpie (Margaret River)
www.swoopingmagpie.com.au
Cabernet franc, Chenin blanc, Verdelho

Symphonia (King Valley)
www.sammiranda.com.au
Arneis, Dolcetto, Meunier, Petit manseng, Pinot gris, Saperavi, Savagnin, Tannat, Tempranillo, Viognier

Symphony Hill Wines (Granite Belt)
www.symphonyhill.com.au
Mondeuse, Moscato paradiso, Picpoul, Pinot gris, Tannat, Tempranillo, Verdelho, Viognier

Syrahmi (Heathcote)
Viognier

T

T'Gallant (Mornington Peninsula)
www.visitor.com.au/tgallant.html
Moscato, Pinot gris, Viognier

Taemus Wines (Canberra)
www.taemaswines.com.au
Shiraz Viognier

Tahbilk (Nagambie Lakes)
www.tahbilk.com.au
Cabernet franc, Grenache, Malbec, Marsanne, Mourvedre, Pinot gris, Roussanne, Sangiovese, Tempranillo, Verdelho, Viognier

Tait Wines (Barossa Valley)
www.taitwines.com.au
Grenache

Talijancich (Swan Valley)
www.taliwine.com.au
Chenin blanc, Graciano, Grenache, Muscadelle, Tempranillo, Verdelho

Tall Poppy (Murray Darling)
www.tallpoppywines.com
Petit verdot, Sangiovese, Viognier

Tallarook Wines (Upper Goulburn)
www.tallarook.com
Marsanne, Mourvedre, Roussanne, Viognier

Tallavera Grove Winery (Hunter Valley)
www.tallaveragrove.com.au
Verdelho

Tallis Wine Company (Goulburn Valley)
www.talliswine.com.au
Sangiovese, Viognier

Taltarni (Pyrenees)
www.taltarni.com.au
Pinot gris, Viognier

Talunga (Adelaide Hills)
www.talunga.com.au
Grenache, Nebbiolo, Petit verdot, Sangiovese, Tempranillo

Tamar Ridge (Northern Tasmania)
www.tamarridgewines.com.au
Gewurztraminer, Pinot gris, Savagnin, Viognier

Tamborine Estate Wines (Queensland Coastal)
www.tamborineestate.com.au
Cabernet franc, Malbec, Verdelho

Tamburlaine (Hunter Valley)
www.mywinery.com
Cabernet franc, Chambourcin, Malbec, Marsanne, Petit verdot, Verdelho

Taminick Cellars (Glenrowan)
www.taminickcellars.com.au
Alicante Bouchet, Durif, Trebbiano

WINERIES • T

Tangaratta Estate (Northern Slopes Zone)
www.tangarattavineyards.com.au
Verdelho

Tanglewood Downs (Mornington Peninsula)
www.tanglewood.com
Gewurztraminer

Tanglewood Vines (Blackwood Valley)
www.tanglewoodvines.com
Viognier

Tanjil Wines (Gippsland)
tanjilwines.com
Pinot gris

Tannery Lane (Bendigo)
Nebbiolo, Sangiovese

Tantemaggie (Pemberton)
Verdelho

Tapestry (McLaren Vale)
www.tapestrywines.com.au
Grenache, Verdelho, Viognier

Tar and Roses (Nagambie Lakes)
Nebbiolo, Pinot gris, Sangiovese, Tempranillo

Tarcoola Estate (Geelong)
Muller Thurgau

Tarup Ridge Winery (Strathbogie Ranges)
www.winediva.com.au/taruplodge/#tarup
Pinot gris

Tassell Park Wines (Margaret River)
www.tassellparkwines.com
Chenin blanc

Tatachilla (McLaren Vale)
www.tatachillawinery.com.au
Chenin blanc, Grenache, Malbec, Sangiovese, Viognier

Tatehams Wines (Clare Valley)
Sangiovese

Tawonga Vineyard (Alpine Valleys)
www.tawongavineyard.com
Durif, Marsanne, Verdelho, Viognier

Taylors (Clare Valley)
www.taylorswines.com.au
Crouchen, Gewurztraminer, Pinot gris, Viognier

Te-Aro (Barossa Valley)
www.te-aroestate.com
Grenache, Mourvedre, Pinot gris, Tempranillo

Telgherry (Hunter Valley)
www.telegherry.com.au
Pinot gris, Viognier

Temple Bruer (Langhorne Creek)
www.templebruer.com.au
Chenin blanc, Grenache, Malbec, Petit verdot, Verdelho, Viognier

Tempus Two (Hunter Valley)
www.tempustwo.com.au
Aranel, Arneis, Marsanne, Moscato, Pinot gris, Sangiovese, Tempranillo, Verdelho, Zinfandel

Ten Miles East (Adelaide Hills)
tenmileseast.com
Arneis, Carmenere, Saperavi

Ten Minutes by Tractor (Mornington Peninsula)
www.tenminutesbytractorwineco.com
Pinot gris, Tempranillo

Tenafeate Creek Wines (Adelaide Plains)
www.tenafeatecreekwines.com.au
Grenache, Nebbiolo, Petit verdot, Sangiovese

Terra Felix (Upper Goulburn)
www.terrafelix.com.au
Marsanne, Moscato, Mourvedre, Roussanne, Viognier

Terrel Estate Wines (Riverina)
Tempranillo

188 WINERIES • T

Tertini Wines (Southern Highlands)
www.tertiniwines.com.au
Arneis

Teusner (Barossa Valley)
www.teusner.com.au
Grenache, Mourvedre

The Cups Estate (Mornington Peninsula)
www.thecupsestate.com
Pinot gris

The Deanery Vineyards (Adelaide Hills)
Sangiovese

The Gap (Grampians)
Grenache

The Garden Vineyard (Mornington Peninsula)
Pinot gris

The Grapes of Ross (Barossa Valley)
www.grapesofross.com.au
Grenache, Moscato, Ruby Cabernet

The Grove Vineyard (Margaret River)
www.thegrovevineyard.com.au
Graciano, Tempranillo, Verdelho

The Islander Estate Vineyards (Kangaroo Island)
www.islanderestatevineyards.com.au
Grenache, Malbec, Sangiovese, Viognier

The Lane (Adelaide Hills)
www.thelane.com
Pinot gris, Viognier

The Lily Stirling Range (Great Southern)
www.thelily.com.au
Chenin blanc, Grenache

The Minya Winery (Geelong)
www.lyrebirdridge.com
Gewurztraminer, Grenache

The Natural Wine Company (Swan Valley)
www.naturalwineco.com.au
Chenin blanc, Verdelho

The Old Faithful Estate (McLaren Vale)
Grenache, Mourvedre

The Pawn Wine Company (Langhorne Creek)
www.thepawn.com.au
Petit verdot, Pinot gris, Sangiovese, Tempranillo, Viognier

The Ritual (Peel)
www.theritual.com.au
Grenache, Mourvedre, Viognier

The Silos Estate (Shoalhaven Coast)
www.thesilos.com
Gewurztraminer, Malbec, Verdelho

The Standish Wine Company (Barossa Valley)
www.standishwineco.com
Viognier

The Wanderer (Yarra Valley)
www.wandererwines.com
Gewurztraminer, Moscato

Thomson Brook Wines (Geographe)
www.thomsonbrookwines.com.au
Barbera, Verdelho

Thomson Estate (Riverland)
www.scottscreekwines.com.au
Tempranillo

Thorn-Clarke Wines (Barossa Valley)
www.thornclarkewines.com
Nebbiolo, Petit verdot, Pinot gris

Three Moon Creek (Queensland Zone)
Marsanne, Petit verdot, Verdelho, Viognier

Three Willows Vineyard (Northern Tasmania)
www.threewillowsvineyard.com.au
Baco noir, Pinot gris

WINERIES • T

Tilba Valley (South Coast Zone)
www.tilbavalleywines.com
Chambourcin, Gewurztraminer

Tilbrook Estate (Adelaide Hills)
www.splitrockvineyard.com.au
Pinot gris, Sangiovese

Tiltili Wines (Langhorne Creek)
Viognier

Tim Adams (Clare Valley)
www.timadamswines.com.au
Grenache, Pinot gris, Tempranillo, Zinfandel

Tim Smith Wines (Barossa Valley)
www.timsmithwines.com.au
Grenache, Mourvedre, Viognier

Tin Shed Wines (Eden Valley)
www.tinshedwines.com
Grenache, Mourvedre

Tinonnee Vineyard (Hunter Valley)
www.tinoneewines.com.au
Chambourcin, Durif, Verdelho

Tintara (McLaren Vale)
www.tintara.com.au
Grenache, Sangiovese, Tempranillo

Tintilla Wines (Hunter Valley)
www.tintilla.com
Sangiovese

Tipperary Estate (South Burnett)
www.tipperaryestate.com.au
Verdelho

Tizzana Winery (South Coast Zone)
www.winery.tizzana.com.au
Aleatico, Petit verdot

TK Wines (Adelaide Hills)
www.tkwines.com.au
Gewurztraminer

Tobin Wines (Granite Belt)
tobinwines.com.au
Tempranillo, Verdelho

Tokar Estate (Yarra Valley)
www.tokarestate.com.au
Tempranillo

Tombstone Estate (Western Plains)
Barbera, Sangiovese

Tomich Hill (Adelaide Hills)
www.tomichhill.com.au
Gewurztraminer, Pinot gris

Toms Cap (Gippsland)
www.tomscap.com.au
Gewurztraminer

Toogoolah Wines (Orange)
www.toogoolah.com.au
Pinot gris

Toolangi Vineyard (Yarra Valley)
www.toolangi.com
Pinot gris

Toolleen Vineyard (Heathcote)
Durif

Toowoomba Hills Estate
(Queensland Zone)
Malbec

Toppers Mountain (New England)
www.toppers.com.au
Barbera, Gewurztraminer, Nebbiolo, Tannat, Tempranillo, Touriga, Verdelho

Torambre Wines (Riverland)
www.torambre.com.au
Mourvedre, Verdelho

Torbreck Vintners (Barossa Valley)
www.torbreck.com
Grenache, Marsanne, Mourvedre, Roussanne, Viognier

Torzi Matthews (Eden Valley)
www.torzimatthews.com.au
Sangiovese

Totino Wines (Adelaide Hills)
www.totinowines.com.au
Pinot gris, Sangiovese

Tower Estate (Hunter Valley)
www.towerestatewines.com.au
Barbera, Moscato, Sangiovese, Verdelho

Trahna Rutherglen Wines
(Rutherglen)
Durif, Petit verdot, Tempranillo, Viognier

Trandari (Hilltops)
www.trandariwines.com.au
Nebbiolo

Transylvania Winery (Southern New South Wales Zone)
Gewurztraminer, Muscadelle

Trappers Gully (Mount Barker)
Chenin blanc

Trentham Estate (Murray Darling)
www.trenthamestate.com.au
Colombard, Grenache, Lexia, Moscato, Nebbiolo, Petit verdot, Pinot gris, Ruby Cabernet, Taminga, Tannat, Vermentino, Viognier

Truffle Hill Wines (Pemberton)
www.wineandtruffle.com.au
Cabernet franc

Tscharke (Barossa Valley)
www.glaymondwines.com
Graciano, Montepulciano, Savagnin, Tempranillo, Zinfandel

Tuart Ridge (Peel)
www.tuartridgewines.com
Chenin blanc, Verdelho

Tuck's Ridge (Mornington Peninsula)
www.tucksridge.com.au
Pinot gris, Savagnin

Tulley Wells (Upper Goulburn)
Grenache

Tulloch (Hunter Valley)
www.tulloch.com.au
Marsanne, Petit verdot, Verdelho

Turkey Flat Vineyards (Barossa Valley)
www.turkeyflat.com.au
Dolcetto, Grenache, Marsanne, Mourvedre, Pedro Ximenez, Roussanne

Turners Crossing Vineyard (Bendigo)
www.turnerscrossing.com
Picolit, Viognier

Twelve Acres (Nagambie Lakes)
Cabernet franc

Twelve Staves Wine Company (McLaren Vale)
www.shadowfax.com.au
Grenache

Twin Oaks (Queensland Coastal)
www.twinoaks.com.au
Chambourcin, Verdelho

Two Dorks Estate (Heathcote)
Viognier

Two Dragons Wine (Currency Creek)
www.twodragonsvineyard.com.au
Tempranillo

Two Rivers (Hunter Valley)
www.tworiverswine.com.au
Verdelho

Two Tails Wines (Northern Rivers Zone)
Chambourcin, Gewurztraminer, Jacquez, Ruby Cabernet, Verdelho, Villard blanc

Tyrrells (Hunter Valley)
www.tyrrells.com.au
Gewurztraminer, Malbec, Pinot gris, Trebbiano, Verdelho, Viognier

U

Uleybury Wines (Adelaide Zone)
www.uleybury.com
Grenache, Petit verdot, Sangiovese, Zinfandel

Undercliff (Hunter Valley)
www.undercliff.com.au
Chambourcin

Upper Reach Vineyard (Swan Valley)
www.upperreach.com.au
Chenin blanc, Verdelho

V

Vale Creek Wines (Central Ranges Zone)
www.valecreek.com.au
Arneis, Barbera, Dolcetto, Pinot gris, Sangiovese, Vermentino

Vale Vineyard (Mornington Peninsula)
www.valewines.com.au
Arneis, Durif, Pinot gris, Tempranillo, Verduzzo

Valhalla Wines (Rutherglen)
valhallawines.com.au
Durif, Grenache, Marsanne, Mourvedre, Viognier

Valley Wines (Swan District)
Chenin blanc, Grenache, Pedro Ximenez

Varrenti Wines (Grampians)
Grenache, Sangiovese

Vasarelli (Currency Creek)
Sangiovese

Vasse River Wines (Margaret River)
www.vasseriver.com.au
Verdelho

Velo Wines (Northern Tasmania)
www.velowines.com.au
Pinot gris

Vercoes Vineyard (Hunter Valley)
www.vercoesvineyard.com.au
Verdelho

Veritas (Barossa Valley)
www.veritaswinery.com
Grenache, Mourvedre, Viognier

Verona Vineyard (Hunter Valley)
Verdelho

Veronique (Barossa Valley)
Grenache, Mourvedre

Vicarys (Sydney Basin)
www.vicaryswinery.com.au
Gewurztraminer

Vico (Riverina)
Barbera

Victory Point Wines (Margaret River)
www.victorypointwines.com
Malbec, Petit verdot

Villa d'Esta Vineyard (Northern Rivers Zone)
villadesta.com.au
Chambourcin, Chasselas, Ruby Cabernet, Verdelho

Vinaceous (Various)
www.vinaceous.com.au
Grenache, Tempranillo, Verdelho

Vincognita (McLaren Vale)
www.vincognita.com.au
Gewurztraminer, Viognier, Zinfandel

Vinden Estate (Hunter Valley)
www.vindenestate.com.au
Alicante Bouchet

Vinea Marson (Heathcote)
www.vineamarson.com
Barbera, Nebbiolo, Sangiovese, Viognier

Vineyard 28 (Geographe)
Chenin blanc, Nebbiolo

Vinifera Wines (Mudgee)
www.viniferawines.com.au
Graciano, Grenache, Tempranillo

Vino Italia (Swan Valley)
Chenin blanc, Grenache

Vinrock (McLaren Vale)
www.vinrock.com
Grenache

Vintara (Rutherglen)
www.vintara.com.au
Cinsaut, Dolcetto, Durif, Grenache, Mourvedre, Petit verdot, Sangiovese, Tempranillo, Viognier

Vintina Estate (Mornington Peninsula)
www.splitrockvineyard.com.au
Pinot gris

Virage (Margaret River)
Gewurztraminer, Zinfandel

Virgara Wines (Adelaide Plains)
www.virgarawines.com.au
Alicante Bouchet, Grenache, Malbec, Sangiovese

Voyager Estate (Margaret River)
www.voyagerestate.com.au
Chenin blanc, Viognier

W

W Wine of Mudgee (Mudgee)
www.wwine.com.au
Sangiovese, Viognier

Wagga Wagga Winery (Riverina)
www.waggawaggawinery.com.au
Touriga

Walden Woods Farm (New England)
www.newenglandwines.org.au
Pinot gris

Walla Wines (Big Rivers Zone)
www.wallawines.com.au
Taminga, Tarrango

Wallambah Vale Wines (Northern Rivers Zone)
www.greatlakes.org.au/accom_result1/wallambah-vale-wines
Chambourcin, Verdelho

Wallington Wines (Cowra)
www.wallingtonwines.com.au
Grenache, Mourvedre, Petit verdot, Tempranillo, Viognier

Walsh Family Winery (Perth Hills)
Gewurztraminer

Walter Clappis Wine Co (McLaren Vale)
www.hedonistwines.com.au
Tempranillo, Viognier

Wandering Brook Estate (Peel)
www.wanderingbrookestate.com.au
Chenin blanc, Verdelho

Wandering Lane (Peel)
Zinfandel

Wandin Valley Estate (Hunter Valley)
www.wandinvalley.com.au
Verdelho

Wandoo Farm (Central Western Australian Zone)
Marsanne, Verdelho, Viognier, Zinfandel

Wanted Man (Heathcote)
www.wantedman.com.au
Dolcetto, Marsanne, Viognier

Waratah Vineyard (Queensland Zone)
www.waratahvineyard.com.au
Marsanne, Petit verdot, Verdelho, Viognier

Warburn Estate (Riverina)
www.warburnestate.com.au
Dolcetto, Pinot gris, Verdelho

Warrabilla Wines (Rutherglen)
www.warrabillawines.com.au
Durif, Marsanne

Warraroong Estate (Hunter Valley)
www.warraroongestate.com
Chenin blanc, Malbec, Ruby Cabernet, Verdelho

Warrego (Queensland Zone)
www.warregowines.com.au
Verdelho, Viognier

WINERIES • W

Warrenmang Vineyard (Pyrenees)
www.warrenmang.com.au
Barbera, Dolcetto, Gewurztraminer, Nebbiolo, Sangiovese

Watchbox Wines (Rutherglen)
www.watchboxwines.com.au
Durif, Sangiovese, Tempranillo

Watershed Wines (Margaret River)
www.watershedwines.com.au
Cabernet franc, Viognier, Zinfandel

Waterwheel Wines (Bendigo)
www.waterwheelwine.com
Malbec, Roussanne

Wattle Ridge Wines (Blackwood Valley)
www.wattleridgewines.com.au
Verdelho

Wattlebrook Vineyard (Hunter Valley)
www.wattlebrook.com
Verdelho

Waurn Ponds Estate (Geelong)
www.waurnpondsestate.com.au
Petit verdot, Viognier

Waybourne (Geelong)
Pinot gris, Trebbiano

Wedgetail Ridge Estate (Darling Downs)
www.wedgetailridge.com.au
Durif, Viognier

Wellington (Southern Tasmania)
Pinot gris

Wells Parish Wines (Mudgee)
Verdelho

Welshmans Reef Vineyard (Bendigo)
www.welshmansreef.com
Tempranillo

Wendouree (Clare Valley)
Malbec, Mourvedre

Wenzel Family Wines (Langhorne Creek)
www.langhornewines.com.au
Petit verdot

West Cape Howe Wines (Denmark)
www.wchowe.com.au
Tempranillo, Viognier

Westend Estate (Riverina)
www.westendestate.com
Durif, Gewurztraminer, Moscato, Pinot gris, Saint Macaire, Tempranillo, Viognier

Western Range Wines (Perth Hills)
www.westernrangewines.com.au
Carnelian, Chenin blanc, Grenache, Malbec, Verdelho, Viognier

Westfield (Swan Valley)
www.westfieldwines.com.au
Chenin blanc, Verdelho

Westgate Vineyard (Grampians)
www.westgate vineyard.com.au
Viognier

Westlake Vineyards (Barossa Valley)
www.westlakevineyards.com.au
Petit verdot, Viognier

Whale Coast Wines (Southern Fleurieu)
www.whalecoastwines.com.au
Petit verdot, Tempranillo, Viognier

Whicher Ridge (Geographe)
whicherridge.com.au
Viognier

Whinstone Estate (Mornington Peninsula)
www.whinstone.com.au
Melon de Bourgogne, Muscadelle, Pinot gris

Whiskey Gully Wines (Granite Belt)
www.whiskeygullywines.com.au
Colombard

Whistle Stop Wines (South Burnett)
www.whistlestop.com.au
Verdelho

Whistling Eagle Wines (Heathcote)
whistlingeagle.com
Sangiovese, Viognier

White's Vineyard (Swan Valley)
www.whitesvineyard.com.au
Grenache, Verdelho

White Rock Vineyard (Northern Tasmania)
Pinot gris

Whitehorse Wines (Ballarat)
Muller Thurgau

Whitsend Estate (Yarra Valley)
www.whitsend.com.au
Viognier

Whyworry Wines (New England)
www.whyworrywines.com.au
Gewurztraminer, Pinot gris, Pinotage, Viognier

Wild Broke Wines (Hunter Valley)
Barbera

Wild Cattle Creek Winery (Yarra Valley)
www.wildcattlecreek.com
Pinot gris

Wild Dog Winery (Gippsland)
www.wilddogwinery.com
Cabernet franc

Wildwood (Sunbury)
www.wildwoodvineyards.com.au
Petit verdot, Viognier

Wili-Wilia Winery (Macedon Ranges)
www.wwwinery.com.au
Gewurztraminer

Wilkie Estate (Adelaide Plains)
www.wilkieestatewines.com.au
Verdelho

Willespie (Margaret River)
www.willespie.com.au
Verdelho

Williams Springs Road (Kangaroo Island)
Petit verdot

Willow Bridge Estate (Geographe)
www.willowbridgeestate.com
Tempranillo, Viognier

Wills Domain Vineyard (Margaret River)
www.willsdomain.com.au
Malbec, Petit verdot, Viognier

Willunga 100 Wines (McLaren Vale)
www.willunga100.com
Grenache, Viognier

Wilmot Hills Vineyard (Northern Tasmania)
www.wilmothills.tascom.net
Gamay, Muller Thurgau, Siegerrebe

Wilson Vineyard (Clare Valley)
www.wilsonvineyard.com.au
Gewurztraminer, Tempranillo, Zinfandel

Winbirra Vineyard (Mornington Peninsula)
www.winbirravineyards.com.au
Meunier, Pinot gris, Viognier

Winbourne Wines (Hunter Valley)
www.winbournewines.com
Verdelho

Winchelsea Estate (Geelong)
Pinot gris

Windance Wines (Margaret River)
www.windance.com.au
Chenin blanc

Windemere Wines (Granite Belt)
Sangiovese

Windowrie Estate (Cowra)
www.windowrie.com.au
Petit verdot, Sangiovese, Tempranillo, Verdelho

Windows Margaret River (Margaret River)
www.windowsmargaretriver.com
Chenin blanc

WINERIES • W

Windshaker Ridge (Perth Hills)
www.windshakerwine.com.au
Carnelian, Verdelho

Windsors Edge (Hunter Valley)
www.windsorsedge.com.au
Tempranillo, Tinto Cao, Touriga, Verdelho

Windy Creek Estate (Swan Valley)
www.windycreekestate.com
Chenin blanc, Grenache, Verdelho

Windy Ridge Vineyard (Gippsland)
www.windyridgewinery.com.au
Gewurztraminer, Malbec

Winewood (Granite Belt)
Marsanne, Viognier

Winooka Park (Central Ranges Zone)
Gewurztraminer

Winter Creek Wine (Barossa Valley)
www.wintercreekwine.com.au
Grenache

Winya Wines (Queensland Zone)
www.winyawines.com.au
Malbec

Wirilda Creek (McLaren Vale)
Verdelho

Wirra Wirra (McLaren Vale)
www.wirra.com.au
Arneis, Grenache, Moscato, Mourvedre, Petit verdot, Viognier

Wirruna Estate (North East Victoria)
www.wirrunawines.com
Durif, Marsanne

Wise Wine (Margaret River)
www.wisevineyards.com
Sangiovese, Verdelho, Zinfandel

Witches Falls Winery (Granite Belt)
witchesfalls.com.au
Durif, Fiano, Grenache, Marsanne

Witchmount Estate (Sunbury)
www.witchmount.com.au
Barbera, Nebbiolo, Picolit, Pinot gris, Tempranillo

Wombat Lodge (Margaret River)
www.wombatlodgewines.com.au
Malbec, Petit verdot

Wonbah Estate (Queensland Coastal)
www.wonbahwinery.com
Verdelho

Wonganella Wines (Northern Rivers Zone)
Verdelho

Wood Park (King Valley)
www.woodparkwines.com.au
Pinot gris, Roussanne, Sangiovese, Verdelho, Viognier, Zinfandel

Woodlands (Margaret River)
www.woodlandswines.com
Cabernet franc, Malbec

Woodonga Hill (Hilltops)
Gewurztraminer, Meunier, Touriga

Woodstock (McLaren Vale)
www.woodstockwine.com.au
Barbera, Grenache, Malbec, Petit verdot, Verdelho

Woody Nook (Margaret River)
www.woodynook.com.au
Chenin blanc

Woolybud (Kangaroo Island)
Sangiovese

Woongoroo Estate (Queensland Coastal)
www.westatewine.com
Verdelho

Woop Woop Wines (McLaren Vale)
www.woopwoop.com.au
Marsanne, Roussanne, Verdelho, Viognier

Word of Mouth Wines (Orange)
www.wordofmouthwines.com.au
Pinot gris

Wordsworth Wines (Geographe)
www.wordsworthwines.com.au
Chenin blanc, Petit verdot, Verdelho, Zinfandel

WINERIES • Y

Wovenfield (Geographe)
www.wovenfield.com
Tempranillo, Viognier, Verdelho

Wrattonbully Vineyards (Wrattonbully)
Marsanne, Tempranillo, Viognier

Wright Family Wines (Hunter Valley)
www.mistyglen.com.au
Chambourcin

Wright Robinson of Glencoe (New England)
www.wrightwine.com
Pinot gris

Wroxton Wines (Eden Valley)
www.wroxton.com.au
Gewurztraminer

Wyuna Park Vineyard (Geelong)
www.wyunapark.com.au
Pinot gris

Y

Yabby Lake Winery (Mornington Peninsula)
www.yabbylake.com
Pinot gris

Yacca Paddock Vineyards (Adelaide Hills)
www.yaccapaddock.com
Arneis, Dolcetto, Durif, Tannat, Tempranillo

Yaldara (Barossa Valley)
www.yaldara.com.au
Grenache, Mourvedre, Petit v erdot

Yalumba Wine Company (Barossa Valley)
www.yalumba.com
Cienna, Grenache, Marsanne, Mourvedre, Nebbiolo, Petit verdot, Sangiovese, Tempranillo, Vermentino, Viognier

Yandoit Hill Winery (Bendigo)
Arneis, Barbera, Nebbiolo

Yangarra Estate (McLaren Vale)
www.yangarra.com
Grenache, Mourvedre, Roussanne, Viognier

Yanmah Ridge (Manjimup)
www.yanmahridge.com.au
Sangiovese

Yarra Burn (Yarra Valley)
www.hardywines.com.au
Pinot gris

Yarra Glen (Yarra Valley)
www.yarraglenwines.com
Marsanne

Yarra Ridge (Yarra Valley)
www.beringerblass.com.au
Pinot gris

Yarra Yarra (Yarra Valley)
www.yarrayarravineyard.com.au
Viognier

Yarra Yering (Yarra Valley)
www.yarrayering.com
Sangiovese, Viognier

Yarraloch (Yarra Valley)
www.yarraloch.com.au
Arneis, Viognier9999

Yarraman Estate (Hunter Valley)
www.yarramanestate.com
Chambourcin, Gewurztraminer

Yarran (Riverina)
www.yarranwines.com.au
Chenin blanc, Petit verdot

Yarrawa Estate (Shoalhaven Coast)
www.yarrawaestate.com
Chambourcin, Verdelho

Yarrawood (Yarra Valley)
www.yarrawood.com.au
Verdelho

Yarrh Wines (Canberra)
www.yarrhwines.com.au
Sangiovese

Yass Valley Wines (Canberra)
www.yassvalleywines.com.au
Barbera, Gewurztraminer, Verdelho

Yaxley Estate (Southern Tasmania)
www.yaxleyestate.com
Pinot gris

Yelland and Papps (Barossa Valley)
www.yellandandpapps.com
Grenache

Yellymong (Swan Hill)
Pinot gris

Yengari Wine Company
(Beechworth)
www.yengari.com
Viognier

Yering Station (Yarra Valley)
www.yering.com
Marsanne, Nebbiolo, Pinot gris, Viognier

Yeringberg (Yarra Valley)
www.yeringberg.com
Marsanne, Roussanne

Yokain Vineyard Estate (Geographe)
www.yokain.com.au
Verdelho, Zinfandel

Z

Zilzie Wines (Murray Darling)
www.zilzie.com
Moscato, Pinot gris, Petit verdot, Sangiovese, Tempranillo, Viognier

Zitta Wines (Barossa Valley)
www.zitta.com.au
Grenache, Mourvedre

Zonte's Footstep (Langhorne Creek)
www.zontesfootstep.com.au
Arneis, Barbera, Dolcetto, Graciano, Grenache, Lagrein, Malbec, Mourvedre, Petit verdot, Pinot gris, Roussanne, Tannat, Teroldego, Tempranillo, Sangiovese, Verdelho, Vermentino, Viognier, Zinfandel

References

BOOKS AND WEBSITES

The major references used in this book are listed below. They are coded in the Varieties chapter

C - Clarke, Oz & Margaret Rand *Grapes and Vines: a Comprehensive Guide to Varieties and Flavours* (London: Websters International Publishers 2003)

D - Delong, Steve & Deborah De Long *Wine Grape Varietal Table: Wine and Grape Indexes* (London: De Long Wine Info 2004)

H - Halliday, James *Varietal Wines (*Sydney: HarperCollins 2004)

K.- Kerridge, George and Allan Antcliff *Wine Grape Varieties Revised edition* (Melbourne: CSIRO 1999*)*

O - Robinson, Jancis *Oxford Companion to Wine, 3rd edition (*Oxford: Oxford University Press 2006)

Further references

Halliday, James *Australian Wine Companion, Various Editions* (Melbourne: Hardie Grant Books 2005-2009)

Halliday, James *James Halliday's Wine Atlas of Australia (*Melbourne: Hardie Grant Books 2006)

Johnson, Hugh & Jancis Robinson *World Atlas of Wines, 5th Edition (*London: Mitchell Beazley 2001)

MacNeil, Karen *The Wine Bible (*New York: Workman Publishing 2001)

Websites

De Long's Wine Info makers of the Wine Grape Varietal Table and Maps http://www.delongwine.com/

Genxy Wines A collection of tasting notes from Melbourne based wine writer Graham Hastings http://genxywines.com/

Vinodiversity This book originated as a website: www.vinodiversity.com

Wine Diva A comprehensive directory of the Australian wine Industry including listings of wineries, regions, wines and varieties www.winediva.com.au

Wine Biz A Business Portal by the publisher Winetitles www.winebiz.com.au

Made in the USA
Lexington, KY
18 March 2012